Social Media Analytics for User Behavior Modeling

A Task Heterogeneity Perspective

Data-Enabled Engineering Series
Series Editor: Nong Ye, Arizona State University, Phoenix, USA

Data Mining
Theories, Algorithms, and Examples
Nong Ye

Convolutional Neural Networks in Visual Computing
A Concise Guide
Ragav Venkatesan and Baoxin Li

Software-Defined Networking and Security
From Theory to Practice
Dijiang Huang, Ankur Chowdhary, and Sandeep Pisharody

Social Media Analytics for User Behavior Modeling
A Task Heterogeneity Perspective
Arun Reddy Nelakurthi and Jingrui He

Attribute-Based Encryption and Access Control
Dijiang Huang, Qiuziang Dong, and Yan Zhu

For more information about this series, please visit: www.crcpress.com/
Data-Enabled-Engineering/book-series/CRCDATENAENG

Social Media Analytics for User Behavior Modeling

A Task Heterogeneity Perspective

Arun Reddy Nelakurthi
Jingrui He

CRC Press
Taylor & Francis Group
Boca Raton London New York

CRC Press is an imprint of the
Taylor & Francis Group, an **informa** business

CRC Press
Taylor & Francis Group
6000 Broken Sound Parkway NW, Suite 300
Boca Raton, FL 33487-2742

International Standard Book Number-13: 978-0-367-21158-5 (Hardback)

Visit the Taylor & Francis Web site at
http://www.taylorandfrancis.com

and the CRC Press Web site at
http://www.crcpress.com

Printed in the United Kingdom
by Henry Ling Limited

Contents

Preface

User-generated social media content provides an excellent opportunity to mine data of interest and helps in developing functional data-driven applications. The rise in the number of healthcare-related social media platforms and the volume of healthcare knowledge available online in the last decade have resulted in increased social media usage for personal healthcare. In the United States, nearly ninety percent of adults, in the age group 50-75, have used social media to seek and share health information. Motivated by the growth of social media usage, this book focuses on healthcare-related applications, studies various challenges posed by social media data, and addresses them through novel and effective machine learning algorithms.

The content presented in this book will be of great interest to students and researchers in the field of Machine Learning with applications to Social Media and Healthcare. This book assumes the reader has sufficient understanding of the theory of machine learning and linear algebra. The reader is suggested to refer to the relevant literature cited for better understanding of proposed frameworks. We are also grateful to collaborators, Dr. Curtiss B. Cook, Dr. Ross Maciejewski and Dr. Hanghang Tong for their valuable feedback and suggestions. We also thank all the colleagues at the STAR and DATA labs. Thanks to Dawei Zhou, Yao Zhou, Xu Liu, Xue Hu, Jun Wu, Lecheng Zheng, Pei Yang, Liangyue Li, Chen Chen, Xing Su, Si Zhang, Boxin Du, Qinghai Zhou, Jian Jian Kang, Zhe Xu, Scott Freitas, Haichao Yu, Ruiyue Peng, Rongyu Lin and Xiaoyu Zhang for their support.

Acknowledgment

This work is partially supported by the United States National Science Foundation under Grant No. IIS-1552654, and Grant No. IIS-1813464, the U.S. Department of Homeland Security under Grant Award Number 17STQAC00001-02-00, and an IBM Faculty Award. The views and conclusions are those of the authors and should not be interpreted as necessarily representing the official policies, either expressed or implied, of the funding agencies or the government.

Authors

Arun Reddy Nelakurthi is a senior engineer in Machine Learning Research at Samsung Research America, Mountain View, California. He received his PhD in Machine Learning from Arizona State University in 2019. His research focuses on heterogeneous machine learning, transfer learning, user modeling and semi-supervised learning, with applications in social network analysis, social media analysis and healthcare informatics. He has served on the program committee for The Conference on Information and Knowledge Management (CIKM) and The Pacific-Asia Conference on Knowledge Discovery and Data Mining (PAKDD). He also worked as a reviewer for IEEE Transactions on Knowledge and Data Engineering (TKDE), Data Mining and Knowledge Discovery (DMKD) and IEEE Transactions on Neural Networks and Learning Systems (TNNLS) journals.

Jingrui He is an associate professor in the School of Information Sciences at the University of Illinois at Urbana-Champaign. She received her PhD in machine learning from Carnegie Mellon University in 2010. Her research focuses on heterogeneous machine learning, rare category analysis, active learning and semi-supervised learning, with applications in social network analysis, healthcare, and manufacturing processes. Dr. He is the recipient of the 2016 NSF CAREER Award and a three-time recipient of the IBM Faculty Award, in 2018, 2015 and 2014 respectively. She was selected for an IJCAI 2017 Early Career Spotlight, and was invited to the 24th CNSF Capitol Hill Science Exhibition. Dr. He has published more than 90 refereed articles, and is the author of the book, Analysis of Rare Categories (Springer-Verlag, 2011). Her papers have been selected as "Best of the Conference" by ICDM 2016, ICDM 2010, and SDM 2010. She has served on the senior program committee/program committee for Knowledge Discovery and Data Mining (KDD), International Joint Conference on Artificial Intelligence (IJCAI), Association for the Advancement of Artificial Intelligence (AAAI), SIAM International Conference on Data Mining (SDM), and International Conference on Machine Learning (ICML).

Contributors

Curtiss B. Cook
Mayo Clinic
Scottsdale, Arizona

Ross Maciejewski
Arizona State University
Tempe, Arizona

Hanghang Tong
University of Illinois
Urbana-Champaign, Illinois

1 Introduction

In recent years, social media has gained significant popularity and become an essential medium of communication. According to a survey, about 88% of the public in the United States use some form of social media, a 53% growth in the last decade. Also, the average number of accounts per user has increased from two in 2012 to seven in 2016 [Pew Research Center, d]. The rise in social media usage both vertically in terms of the number of users by platform and horizontally in terms of the number of platforms per user has led to a data explosion. Popular social media platforms like Facebook, Instagram and Twitter manage tens of petabytes of information with daily data flows of hundreds of terabytes and a continually expanding userbase [Pew Research Center, c]. Such huge volumes of user-generated content provide an excellent opportunity to mine data of interest. We can, thus, look for valuable nuggets of information by applying diverse search (information retrieval) and mining techniques (data mining, text mining, web mining, opinion mining).

User-generated content is diverse based on the need the social media platform caters to. Per one survey, amongst those who use social media roughly 67% stated staying in touch with current friends and family as a major reason, while 17% felt social media enabled them to connect with friends they have lost touch with [Pew Research Center, a]. Other research indicated about 67% of the United States population use social media to stay updated on the latest news and seniors are driving that number up [Pew Research Center, b].

Social media usage has also seen a spike when it comes to personal healthcare. In the United States, nearly 90% of adults, in the age group 50-75, have used social media to seek and share health information [Tennant et al., 2015a]. Research demonstrates that online social support programs like health care forums and social media websites (e.g. Facebook and Twitter) can help patients gain knowledge about their diseases and cope better with their daily management routine [Petrovski et al., 2015]. Effectively mining information from these healthcare-related social media platforms can, thus, have a wide range of applications resulting in improved healthcare. For example, healthcare social networks can connect patients suffering from major chronic diseases such as diabetes mellitus, with physicians as well as other patients. Compared to generic social networks such as Twitter and Facebook, disease-specific social networks (e.g., TuDiabetes[1] and DiabetesSisters[2]) have a greater concentration of patients with similar conditions and relevant resources. However, when it comes to such social networks, the patient is more likely to stick to a single social network, and would rarely look at other networks, thus limiting their access to online resources, especially patients with similar questions and concerns. Identifying patient

---

[1] http://www.tudiabetes.org/

[2] https://diabetessisters.org/

groups with similar conditions can help connect patients across networks, thereby opening doors for knowledge sharing to help the community as a whole. Additionally, in a world of "fake news", a lot of health information is misrepresented and therefore calls for authenticity. Motivated by the immense scope of leveraging social media information for healthcare and addressing underlying challenges with usage and reliability, in this book we explore answers to the following questions:

- Can social media serve as a platform for improved healthcare? Specifically, why would patients leverage social media and how would it impact their healthcare? Would it equip them to make better choices? And finally, does it help in communicating effectively with doctors and health care providers?
- How can we efficiently learn and build algorithms to mine knowledge from these healthcare-dedicated social networks? What are the challenges involved?
- Finally, how can we provide meaningful explanations to justify the behavior of algorithms and learning methods?

Unlike traditional mining settings where data is considered to be homogeneous for most mining tasks, user-generated social media data is intrinsically heterogeneous and thus poses a set of challenges. It can be both structured (ratings, tags, links) as well as unstructured (text, audio, video). Similar health-related social media websites that cater to users from different geographical locations can suffer from a distributional shift in user-generated data, either features or class labels. This shift could also be due to user bias or personal preferences. Transfer learning addresses the problem of distribution shift in data [Pan and Yang, 2010]. In particular, task heterogeneity is reflected in inconsistent user behaviors across social media platforms, similar actors across social networks, etc. Therefore, in this book work, we aim to design efficient models and tools to help us leverage and learn from data heterogeneity in real-world scenarios that help in improving healthcare.

In scenarios where parts of data in one social network are hidden, missing or not available, leveraging it partially for mining is very challenging and has not been well studied. Motivated by the applications of task heterogeneity, in this book, we present our work on techniques for addressing task heterogeneity and the underlying challenges in social media analytics.

In lieu of the above questions and challenges for this research, three main research directions have been investigated:

D1. **Social media in healthcare**: To study the real-world impact of social media as a source to seek and offer support to patients with chronic health conditions.

D2. **Learning from task heterogeneity**: To propose various models and algorithms to learn and model user behaviors on social media platforms, to identify similar actors across social networks, to adapt and leverage information from existing black-box models to improve classification accuracy under domain adaptation settings.

D3. **Model explainability**: To provide interpretable explanations for heterogeneous predictive models in the presence of task heterogeneity.

The book is organized as follows. The related work, Chapter 2 discusses existing research and how the proposed methods differ from it. Chapter 3 discusses the impact of social media on patients with diabetes mellitus. Chapter 4 presents algorithms and models to learn from task heterogeneity in social media. Chapter 5 discusses methods to explain task heterogeneity. Finally, Chapter 6 concludes our research.

2 Literature Survey

Since 2004, the growth of social media has been near exponential [We Are Social]. According to a survey, about 88% of the public in the United States use some form of social media, a 53% growth in the last decade [Pew Research Center, d]. This growth in social media usage led to an information explosion. Mining valuable nuggets of data from such information generated through social media has immense applications [Zafarani et al., 2014]. Machine learning techniques have been widely adopted to mine and analyze the large social media data to address many real-world problems. Mining from social media platforms has many applications, (1) Event detection - Social networks enable users to freely communicate with each other and share their recent news, ongoing activities or views about different topics. As a result, they can be seen as a potentially viable source of information to understand the current emerging topics/events [Nurwidyantoro and Winarko, 2013]; (2) Community detection - identifying communities on social networks, how they evolve, and evaluating identified communities, often without ground truth [Zafarani et al., 2014]; (3) Recommendation in social media - recommending friends or items on social media sites [Ricci et al., 2011]; (4) Sentiment and opinion mining - identifying collectively subjective information, e.g. positive and negative, from social media data [Liu, 2012]; (5) Network embedding - assigning nodes in a network to low-dimensional representations and effectively preserving the network structure [Cui et al., 2017].

As mentioned earlier in the introduction chapter, the intrinsic property of data heterogeneity in social media data poses a set of challenges. In this chapter, we present the existing research on handling data heterogeneity and study the impact of social media. In this chapter, we present existing work on impact of social media and its implications in Section 2.1, Section 2.2 presents existing research addressing data heterogeneity with a focus on task heterogeneity. Finally, we discuss the existing research on explaining models under task heterogeneity in Section 2.3.

2.1 IMPACT OF SOCIAL MEDIA

The growing popularity in the usage of social media platforms and applications has an impact on the individuals and society as a whole [Bishop, 2017]. These platforms have revolutionized the way we view ourselves, the way we see others and the way we interact with the world around us. Social media has many positive implications. Khurana [2015] studied the impact of social networking sites on the youth; it was shown that social media enables connecting with people all across the globe by not hampering their work hours and schedules and it also helps in education. Hudson and Thal [2013] studied the impact of social media on the consumer decision process and its implications for tourism marketing. Pew Research Center [b], showed that about 67% of the United States population uses social media to stay updated on the latest news. Also, the use of social media in politics including Twitter, Facebook, and

YouTube has dramatically changed the way campaigns are run and how Americans interact with their elected officials [Bonilla and Rosa, 2015]. Social media usage has also seen a spike when it comes to personal healthcare. Tennant et al. [2015b] showed that nearly 90% of adults who use the internet and social media platforms like Facebook and Twitter used these platforms to find and share healthcare information. With a lot of growing interest and immense benefits from healthcare applications to society, we are motivated to work on addressing challenges in healthcare-related social media platforms.

Research demonstrates that online social support programs like health care forums and social media websites (e.g. Facebook and Twitter) can help patients gain knowledge about their diseases and cope better with their daily management routine [Petrovski et al., 2015]. Patel et al. [2015] studied the impact of social networks on perceived social support (e.g., of patients with chronic diseases). Researchers also studied how social media users gather and exchange health-related information and share personal experiences [Naslund et al., 2016, Shepherd et al., 2015]. Fung et al. [2016] researched the spread of misinformation about disease outbreaks to inform public health communication strategies.

2.2 HETEROGENEOUS LEARNING AND SOCIAL MEDIA

Mining from healthcare-related social media platforms is challenging. The key to building applications from social media data is user-behavior modeling. Social media data is intrinsically heterogeneous - generated by users from different demographical locations, who speak different languages and have different cultural backgrounds. This makes user-behavior modeling under heterogeneity very challenging. Further, to mine across multiple social media platforms, the likelihood of the same user having multiple accounts is very low. Often they stick to one or two platforms that are popular based on the geographical location or demographics. To efficiently design applications that serve across multiple platforms, it is essential to identify similar users across the networks. Finally, it is very costly to collect labels for data from multiple platforms. A more practical approach would be to leverage knowledge from one platform to another. Motivated by this we identified three major problems: (1) modeling user-behavior; (2) identifying similar actors and (3) adapting to new domains. In this section, we discuss existing research on each of the problems.

In traditional machine learning models, it is considered that the training data on which the model is trained has similar data distributions to the data at the test time. Due to data heterogeneity and the dynamic nature of social media platforms, it is not possible to use traditional machine learning models. In the past, researchers have addressed these issues through a new branch of machine learning called Transfer Learning. In transfer learning, given data from the source domain and target domain, models are trained on a source domain and the underlying knowledge is transferred to target domain [Pan and Yang, 2010]. Different supervised, unsupervised and semi-supervised methods have been proposed for a wide variety of applications such as image classification [Tan et al., 2015], WiFi-localization on time variant data [Pan et al., 2008], and web document classification [He et al., 2009, Pan et al., 2010].

Transfer learning is broadly classified into inductive, transductive and unsupervised transfer learning [Pan and Yang, 2010]. In inductive transfer learning, the distributions of the data in the source domain and target domain are considered to be similar, but the machine learning task varies from the source domain to the target domain. Self-taught learning [Raina et al., 2007] and multi-task learning [Zhang and Yang, 2017] are a few examples of inductive transfer learning, whereas in transductive transfer learning, the tasks are the same but the data distributions vary from the source domain to the target domain. Domain adaptation based methods [Jiang, 2008], correcting co-variate bias, cross-domain sentiment classification [Blitzer et al., 2007] and cross-domain recommendation are a few examples of transductive learning. In unsupervised transfer learning, the labels in the source domain and target domain are not observable. Self-taught clustering (STC) [Dai et al., 2008] and transferred discriminative analysis (TDA) [Wang et al., 2008] algorithms are proposed to transfer clustering and transfer dimensionality reduction problems, respectively. Given the heterogeneous nature of the data on social media platforms, we are interested in transductive learning in this research work.

2.2.1   TRANSDUCTIVE TRANSFER LEARNING

In transductive transfer learning, the data distributions vary across the source and target domains, but the learning task, sentiment analysis, is the same in both the domains. Sentiment classification in a cross-domain set up is a well-studied problem. For example, structural correspondence learning (SCL) generates a set of pivots using common features in both the source and target domains using mutual information and a set of classifiers on the common features [Blitzer et al., 2007]; spectral feature alignment (SFA) splits the feature space into domain independent features and domain-specific features, then aligns the domain-specific features into unified clusters by using domain independent features as a bridge through spectral feature clustering [Pan et al., 2010]; transfer component analysis (TCA) utilizes both the shared and the mapped domain-specific topics to span a new shared feature space for knowledge transfer [Li et al., 2012]; the labeled-unlabeled-feature tripartite graph-based approach called TRITER was proposed to transfer sentiment knowledge from labeled examples in both the source and target domains to unlabeled examples in the target domain He et al. [2009].

Prior research has shown that user information combined with linguistic features improved sentiment classification. Li et al. [2014] proposed a user-item based topic model which can simultaneously utilize the textual topic and latent user-item factors for sentiment analysis; Tang et al. [2015b] incorporated user- and product-level information using vector space models into a neural network approach for document-level sentiment classification. Motivated by prior work which demonstrated the usefulness of user information in single-domain sentiment classification, we propose *U-Cross* to explicitly model the user behaviors by borrowing information from the source domain to help construct the prediction model in the target domain. Tan et al. [2011] used a factor-graph model for user labels in a transductive learning setting for a short-text sentiment classification task. It is likely that the user behavior can vary across the source and target domains; if not handled well it can lead to the negative transfer of knowledge. Our work on cross-domain sentiment classification varies from Tan et al. [2011] as we carefully model the user behavior based on the relatedness between the source and target domains, which prevents the 'negative transfer'.

2.2.2 SOURCE-FREE TRANSFER LEARNING

Source-free transfer learning is a special case of transductive transfer learning, where there is limited to no knowledge of labeled examples and also the feature distribution of the examples from one or more source domains. Yang et al. [2007] proposed the Adaptive-SVM framework where the goal is to learn the target classification function by adapting the pre-trained classifiers to the labeled examples in the target domain. Duan et al. [2012] and Xiang et al. [2011] proposed the variants of the "Domain Adaptation Machine" (DAM) to learn the target classification function. They assume that there exists multiple source classifiers (black box), and access to a few labeled examples and all the unlabeled examples in the target domain. They extend the Adaptive SVM by introducing a data dependent regularizer on all the examples in the target domain and the labeled examples in the target domain. Our work is significantly different from the DAM, as we consider only one off-the-shelf classifier compared to multiple SVM classifiers used in DAM and also provide a drift correction framework to adapt the off-the-shelf classifier to labeled examples. Lu et al. [2014] proposed a source domain free approach by leveraging the information from existing knowledge sources like WWW or Wikipedia. They build a large label knowledge base with 50,000 category pairs and train classifiers for each of the category pair. The goal is to compute the latent features on the labels, which is further used to compute the target labels from unlabeled examples. The problem of "Source-free transfer learning" in Lu et al. [2014] and Xiang et al. [2011] is different from the problem of off-the-shelf classifier adaptation; instead of building a knowledge base we simply make use of an existing off-the-shelf black-box classifier to improve accuracy on the set of unlabeled examples. In the paper Chidlovskii et al. [2016], authors consider three different scenarios, (1) The parameters of the source classifiers are known; (2) Source classifiers as a black box; (3) Class distribution of the source classifier is known. The case 2 is very relevant to our work. They employ marginalized denoising auto-encoders to denoise the source classifier labels using unlabeled data in the target domain. Our approach is semi-supervised and leverages the similarity between the examples which varies from Chidlovskii et al. [2016] as it is an unsupervised setting.

2.2.3 IDENTIFYING SIMILAR ACTORS ACROSS NETWORKS

Identifying similar actors across networks can be considered as a cross-network link prediction problem. Link prediction is a widely studied problem in the field of social network analysis [Liben-Nowell and Kleinberg, 2007, Al Hasan and Zaki, 2011, Wang et al., 2015]. Link prediction can be broadly classified into two types: (1) Classical link prediction which aims at predicting the missing links in a given social network [Al Hasan et al., 2006, Fortunato, 2010]; (2) Cross network link prediction that recommends the links across two or more social networks. Tang et al. [2012] modeled users as a feature vector within-domain and cross-domain topic distributions, and used it to learn associations between users across the source and target domains. Kong et al. [2013] suggested a multi-network anchoring algorithm to discover the correspondence between accounts of the same user in multiple networks. Zhang et al. [2015] proposed an energy-based framework COSNET for cross-network link prediction in heterogeneous networks. Our problem of network link prediction differs with previous cross network link prediction problems, as we recommend links between similar actors across social networks.

Non-negative matrix factorization (NMF) is widely used for co-clustering problems. Li and Ding [2006] demonstrated a NMF framework for document-word co-clustering. Cai et al. [2011] improved the framework proposed by Li and Ding [2006] by adding a graph regularizer which captures geometric information embedded in the data. Gu et al. [2011] proposed an orthogonal framework to fix scaling problem in Cai et al. [2011]. Wang et al. [2015] proposed an NMF based Dual Knowledge Transfer approach for cross-language Web page classification. Our approach differs from previous works as we jointly factor user-keyword matrices from multiple social networks to learn latent features on the combined set of keywords from all the social networks and users from each social network. Chakraborty and Sycara [2015] proposed a constrained NMF framework for community detection in social networks which is closely related to our work. Our problem is different from the community detection problem, which finds communities of closely related users inside a social network.

2.3 EXPLAINING TASK HETEROGENEITY

Machine learning, today is being used for a wide range of practical decision-making applications. In most cases, the models were considered as black boxes and the decisions made by the system are not explainable. With an increase in research on building more complex models like Deep Neural Networks for improving model performance, there is a need to build techniques to explain the complex models. There are mainly two approaches for explaining the model's performance, via relevant features or through the importance of training samples. The first approach examines the importance of different features to model predictions. To work for any complicated model, LIME Ribeiro et al. was proposed as a model-agnostic explanation model to learn an interpretable model locally around the prediction for a specific test sample. In some cases, the features may have an indirect influence on the model prediction via other related features. Such indirect influence can be quantified based on differential analysis of feature influence before and after obscuring the feature influence on the model outcome [Adler et al., 2018]. The second popular approach to model interpretability is to generate explanations by understanding the influence of training examples. Influence functions Koh and Liang [2017], as a classic technique from robust statistics, was used to trace a model's prediction through the learning algorithm back to its training data. The key idea is to compute the change of the loss at a test sample should a training example be up-weighted by some small ε. A graph signal process has also been used for influential sample analysis where the influence metric is used as a function at the nodes in the data graph [Anirudh et al., 2017]. The most influential samples would be those critical to the recovery of high-frequency components of the function. While most of the existing work on interpretability and model explanation looked into machine learning models in general, we focus on explaining the transfer learning models.

Various methods have been recently proposed to explain such complex models. Ribeiro et al. proposed a model-agnostic framework that can identify the important features for classification. Tolomei et al. [2017] proposed a technique that exploits the internals of a tree-based ensemble classier to offer recommendations for transforming true negative instances into positively predicted ones. Lundberg and Lee [2017] proposed a unified framework for interpreting predictions, SHAP (Shapley Additive exPlanations). The SHAP framework assigns each feature an importance value for a particular prediction. Also, they demonstrated that SHAP value estimation methods are better aligned with human intuition as measured by user studies and more effectually discriminate among model output classes than several existing methods. Koh and Liang [2017] used the classic technique from robust statistics, influence functions, to identify the set of examples that influence the classification and use them to explain the models. Most of the work in the past has focused only on regular machine learning settings. We would like to work on model explainability in transfer learning settings. The data in transfer learning settings are inherently inhomogeneous, leading to differences in the feature and class distribution making the model explainability more challenging in this case.

3 Social Media for Diabetes Management

The emergence of the internet has made it possible for people to go online to seek answers to their health related questions. Many patients use the Internet to find and join communities of individuals with similar health conditions in order to share information, and to provide and receive advice on management. However, little data is available on whether patients with diabetes mellitus (DM) utilize social media. The aim of our study was to investigate the likelihood of patients utilizing social media to offer or seek support from others for their diabetes management.

Diabetes mellitus (DM) is a chronic illness that can be effectively managed through physical activity, healthy dietary habits, and the appropriate and timely use of pharmacotherapies to lower blood glucose levels. Research has demonstrated that online social support programs like health care forums and social media websites (e.g. Facebook and Twitter) can help patients gain knowledge about DM and cope better with their daily management routine [Petrovski et al., 2015]. Such platforms allow patients to share personal clinical information, request disease-specific advice, and even receive the emotional support that they need for diabetes management and self-care [Greene et al., 2011].

Most of the research on the influence of social media on DM care has focused on widely used platforms without a clear focus on a specific disease [Petrovski et al., 2015, Ravert et al., 2004]. Research has shown that such generic social media platforms have lots of promotional activity and personal data collection but no checks for authenticity. On the other hand, most of the diabetes-specific social media platforms are moderated and enforce measures on patient privacy. Often moderated platforms ensure authenticity and correctness in the information delivered to users. Another study of preferences for online DM support found that adults generally preferred professionally moderated discussions. So we focus our study on diabetes-specific social media platforms. Little is known about specific reasons why individuals with DM utilize social media, or if use of social media is associated with DM specific behaviors. In order to answer these questions, and to gain valuable insight on social media use, the aim of this study was to assess the behaviors of individuals with DM who either offered or sought information on diabetes specific social media websites.

3.1 METHODOLOGY

Information regarding DM was collected through an anonymous web-based survey. The data were collected through a self-administered Qualtrics platform, advertised through DM dedicated healthcare forums. As part of the survey, participants were asked to answer questions broadly classified as 1) demographic information,

2) diabetes specific information (e.g., diabetes diagnosis, most recent HbA1c value), 3) nature and frequency of diabetes-specific social networking site usage, and 4) dietary habits and diabetes self-care activities [Toobert et al., 2000]. Participation was anonymous and required individuals to be at least 18 years old. Once individuals read a brief online consent document and agreed to participate, they proceeded with the survey. No compensation was offered for participation. The study was approved by the Baruch University Integrated Institutional Review Board (IRB# 2015-0767).

The demographic, DM-specific and social media usage information were analyzed using descriptive statistics. Associations between the use of a DM-specific social media website and DM-related behaviors were analyzed using correlation analysis with Pearson's correlation coefficient. The responses to the survey questions (Figure 1) on social media website usage contain both ordinal and categorical Likert scale based responses. For convenience the responses were converted into a numerical scale with order and rank preserved. Questions with ordinal responses (e.g. Questions 3-5) were represented using a numerical [1-8] scale as follows: 1: Less than once a month; 2: Once a month; 3: 2-3 times a month; 4: Once a week; 5: 2-3 times a week; 6: 4-5 times a week; 7: Once a day; 8: More than once a day. Similarly, for questions where responses utilized a Likert scale (e.g. Questions 6-9), the responses were represented using a numerical [1-5] scale as follows: 1: Not at all likely; 2: Slightly likely; 3: Moderately likely; 4: Very likely; 5: Most likely. For analysis on descriptive statistics, Likert scale based questions were split into two groups: 1) Not at all likely - Moderately likely 2) Very likely – Most likely. Similarly, ordinal responses were split into two groups: 1) Less than once a month - once a week 2) 2-3 times a week - more than once a day. Questions and their responses were assumed independent of each other. Tests of significance were based on alpha of 0.05.

3.2   RESULTS

A total of 45 participants (mean age $= 57 \pm 13.4$ years) from the United States and United Kingdom combined submitted their responses (Table 1). Among the 45 participants most were women, identified themselves as white, had attended college (38/45 or 84%), and were retired (25/45 or 56%). The participant pool consisted of a balanced mix of patients who reported they had type 1 DM and type 2 DM. The average self-reported hemoglobin A1c was 7.0% (53 mmol/mol), and 28 participants had a self-reported hemoglobin A1c of less than that value. Patients with DM who monitored their hemoglobin A1c at least once every three months reported having the lowest hemoglobin A1c levels ($r = 0.45, p = 0.01$). There were 50 responses (mean age $= 62 \pm 13.2$ yrs, 26 women). Most (61%) had type 2 DM, were Caucasian (66%), and were on insulin (58%). Mean hemoglobin A1c was $7.46 \pm 1.61\%$.

Analysis showed that most of the respondents (45 out of 50) do not actively utilize social media to offer advice or seek support from others for their diabetes management. Also, 5 patients out of 45 patients reported that they do not access social media but would like to get support from the social media for their diabetes management. Table 3.1 shows the descriptive statistics of the online survey.

Table 3.1

Descriptive Statistics of Diabetes Survey Participants

Characteristic		Mean (SD) or N (%)
Age, (years)		57 (14.3)
Sex, No.	Women	30 (66.7 %)
	Men	15 (33.3 %)
Race	White	45 (100 %)
Diabetes Duration, (years)		16.78 (14.1)
Diabetes Diagnosis	Type 1	21 (46.7 %)
	Type 2	22 (48.9 %)
	Other	2 (4.4 %)
Prescribed insulin for diabetes	Yes	29 (64.4 %)
	No	16 (35.6 %)
Hemoglobin A1C	Mean	7.0% (2.2%)
	< 7.0 %	28
	>= 7.0%	17
Education, No.	Did not complete high school	1
	Completed high school	4
	Completed vocational training	2
	Some college (less than 4 years)	15
	Completed college	15
	Graduate or professional degree	8
Employment Status	Working full-time	8
	Working part-time	4
	Not currently working	7
	Student	1
	Retired	25

Of the DM-specific websites listed in the survey, the ones reportedly visited most were Tudiabetes.org (15 respondents), Diabetesdaily.com (11 respondents) and Diabetes-support.org.uk (11 respondents). When asked when they started visiting their typical DM website, most (32/45 or 71%) indicated they had begun using it more than 12 months prior to the survey. The majority (29/45 or 64%) reported posting a question on that website less than one month ago, and approximately half (21/45 or 47%) post information or advice 2-3 times per week. Figure 3.1 shows the 4 most common reasons respondents indicated they would be more than moderately likely to visit the website were to offer support or encouragement, to share personal experiences, or seek support or advice for themselves.

The authors were interested in exploring the reasons why survey respondents went to a social media website to post information on DM. We considered individuals with DM who frequently (at least 2-3 times a week) log in to the DM-specific social media website and analyzed the top reasons to offer and seek information. The top 3 reasons respondents indicated they would be moderately-extremely likely to visit the website and share the information were to offer support or encouragement to other individuals with DM, to share personal experiences and to offer advice about clinical diabetes care. Similarly, the top 3 reasons to visit the website and seek information

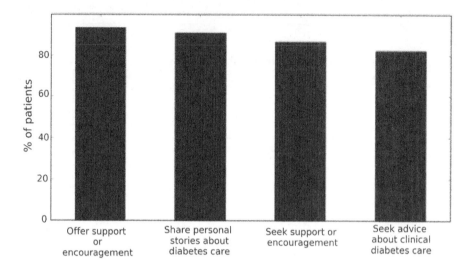

Figure 3.1   Top 4 reasons DM survey participants reportedly visit social media websites.

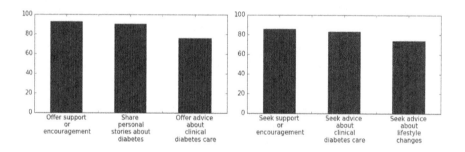

Figure 3.2   3 reasons DM survey participants who login to the social media websites at least 2-3 times a week to (a) post information and (b) seek advice.

is to seek support or encouragement from individuals with DM, to seek advice about clinical diabetes care and life style changes (Figure 3.2).

Next we examined the relationships between social media use and specific DM health behaviors. Significant correlations were observed between respondents offering advice on social media sites and their own self-reported eating and exercise habits over the prior 7 days. Furthermore, using the website to obtain information on lifestyle changes for diabetes management was significantly associated with following that advice and using the website to obtain advice about clinical diabetes care (e.g. blood glucose monitoring) was correlated with perceptions of how helpful

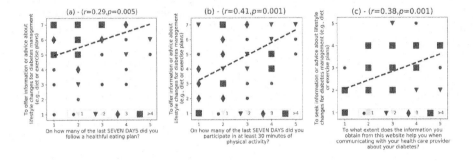

Figure 3.3 Correlation results between information or advice offering and seeking behaviour and other behavioural traits of diabetes mellitus survey participants. The correlation parameters r and p-value are shown in the subfigure title. The symbols represent number of respondents (circle = one response, triangle = two, diamond = three and square = four).

this information was when communicating with health care provider about diabetes (Figure 3.3).

Survey data showed that 39/45 (87%) of respondents were moderately-extremely likely to follow the advice they received from the website about lifestyle changes for DM management, and 38/45 (84%) were moderately-extremely likely to follow the advice received from the website about clinical DM care (e.g., blood sugar monitoring, medications). Moreover, the information obtained from the websites helped about 37/45 (82%) of DM survey participants in communicating with their health care provider about their DM.

3.3 DISCUSSION

We assessed DM-specific social media website use to gain insight into how people with diabetes utilize these resources. Approximately half of the participants reported posting information or advice 2-3 times per week or more and posts were most likely to be about offering support or encouragement to other user, sharing personal stories about diabetes care and seeking support or advice about DM. Results suggested that respondents' frequency of using a social media site was motivated by their own desire to offer support or encouragement to other users with DM. In addition, positive associations were found between likelihood of offering information or advice about lifestyle change and frequency of engaging in lifestyle change for diabetes management. It is possible that these individuals had discovered successful coping or treatment strategies they wished to pass on to the wider DM online community. Study results also showed an association between seeking online information and following management recommendations. Moreover, a majority of participants reported that the information they obtained facilitated communication with their health care providers. Future research that examines the relationship between online posting and diabetes-related self-care behaviors in a longitudinal design would help to clarify the role of website use in diabetes management.

It can be hypothesized that the more active a user is (e.g., offering information or advice about different aspects of DM, sharing personal stories), the more likely they would respond to another user's question or concern. For example, when a website user posts a question on social media, it may lead to a conversation resulting in increased usage of the website for DM management. Such online interactions may help to inform both diabetes management behaviors and communication with health care providers.

There are limitations to the current study. For instance, the sample size is small and the data are self-reported. Moreover, a selection bias exists in that those who responded could have been individuals who were more motivated to manage their DM. An additional limitation is that all study participants were Caucasian. Extending this study to minority and under-served groups to better understand the impact of diabetes online social networks in these populations is an important next step. Finally, a control group of individuals who did not use social media was lacking. Repeating the survey in a clinic based population could help better delineate differences in self-management behavior between social media users and non-users.

Despite the limitations, results of this study showed that individuals with DM who are active in using DM related social media tend to seek and offer information to others. Greater self-reported adherence to DM management behaviors was associated with a greater likelihood of offering information or advice to others on social media about DM care.

3.4   CHALLENGES IN REAL-WORLD APPLICATIONS

Effectively leveraging data mined from healthcare-related social platforms poses a set of challenges. They are as follows:

1. *Identifying similar patients across networks* - most users of these healthcare-related platforms are likely to stick to a single social network, and would rarely look at other networks, thus limiting their access to online resources, especially patients with similar questions and concerns. Identifying patient groups with similar conditions can help connect patients across networks, thereby opening doors for knowledge sharing to help the community as a whole.

2. *Modeling user behavior* - The data in these healthcare-related social media platforms is intrinsically heterogeneous. Patients, healthcare providers, and other users come from different backgrounds and demographics. Also, their dietary habits and methods to cope with chronic diseases might vary from region-to-region. Data heterogeneity brings many challenges in modeling user behavior.

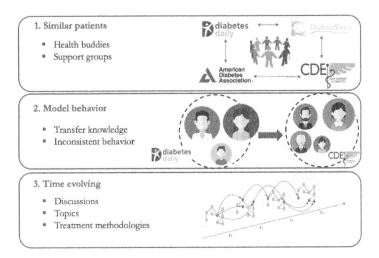

Figure 3.4 Three challenges in leveraging social media data for healthcare applications.

3. *Time evolving* - Healthcare-related platforms are dynamic in nature. The social media data evolves over time, for example, topics of discussion, treatment methodologies, and self-care practices. This calls for applications to be robust and adapt to the change in data distributions.

Efficiently addressing the above challenges can have many applications. Figure 3.4 illustrates various challenges. Identifying similar actors can help in recommending health buddies with similar questions and concerns. Also, in our research, it was seen that social media platforms act as a source for offering and seeking support to individuals. Identifying similar actors can help form support groups, and bring people together with similar interests and demographics. Modeling user behavior through social media data can help identify widely followed chronic care practices and best health-care products. Opinion mining on feature-rich platforms can help in recommending best practices to users on other social media platforms. Knowledge from one healthcare social-media platform can be transferred to other platforms. In the following chapters, we provide algorithms to address the above mentioned challenges along with case-studies on applications related to healthcare.

4 Learning from Task Heterogeneity

4.1 CROSS-DOMAIN USER BEHAVIOR MODELING

Sentiment analysis, or opinion mining, is extremely useful in many real applications such as media monitoring, which allows us to gain an overview of public opinion on stocks, products, movies, politicians, or any other topic that is being discussed. For example, the Obama administration used sentiment analysis to gauge public response to campaign messages during the 2012 presidential election; nonprofit organizations, such as the American Cancer Society, have employed sentiment analysis to gauge feedback on their fundraising programs; and Expedia Canada was able to quickly identify and react to the fact that one of their television advertisements was considered to be annoying[1]. In sentiment analysis, when the target domain (e.g., review articles written in Chinese) has only a limited amount of labeled data, and it is both costly and tedious to collect more labeled information, a common practice is to apply transfer learning, or domain adaptation, which borrows information from a relevant source domain with abundant labeled data (e.g., review articles written in English) to help improve the prediction performance in the target domain [Wan, 2009].

However, most existing transfer learning techniques for sentiment analysis largely overlooked an important factor, the human factor, which is usually associated with the degree of sentiment or opinion making [Blitzer et al., 2007, Pan et al., 2010, Glorot et al., 2011]. In other words, users who are optimistic and positive tend to give high ratings, and vice versa. This bias can also be due to the tendency that users associated with a company or brand usually post positive reviews for their products and negative reviews for their competitors. Therefore, human behavior should be explicitly modeled in transfer learning to effectively leverage such information.

In this work on user behavior modeling, we propose a new graph-based transfer learning approach: User-guided Cross-domain sentiment classification (*U-Cross*). It constructs a user-example-feature tripartite graph, and imposes a set of constraints such that: (1) the sentiment of content generated by the same user is consistent; (2) label information is propagated from the source domain to the target domain via the common keywords; and (3) the subtle language differences between domains are identified by exploiting the label information (abundant from the source domain, and limited from the target domain). This approach is non-parametric and semi-supervised in nature. Furthermore, we address the problem of 'negative transfer' by excluding a set of common users across different domains with known inconsistent

---

[1] http://www.marketingmag.ca/brands/expedia-ca-responds-to-angry-social-media-feedback-with-new-ads-99039

behaviors. To demonstrate the effectiveness of the proposed *U-Cross* approach, we test it on three different datasets of varied sizes, and compare it with state-of-the-art techniques on cross-domain sentiment classification. The following are the major contributions towards modeling user behavior in cross-domain settings:

1. A novel graph-based framework for cross-domain sentiment classification, leveraging user-example-feature relationships.
2. A novel user selection approach to avoid negative transfer through soft-score reweighting, and to gauge the consistency of users across the source and target domains.
3. Extensive experimental analysis to demonstrate the effectiveness of *U-Cross* over state-of-the-art cross-domain sentiment classification approaches.

The rest of the work on user behavior modeling is organized as follows. In Subsection 4.1.1, we introduce our proposed graph-based approach *U-Cross*, its algorithm and proof of its convergence. A special case of the proposed approach is discussed in Subection 4.1.2, which is equivalent to an existing method TRITER [He et al., 2009]. Then we demonstrate the effectiveness of *U-Cross* in Subsection 4.1.3 on multiple real datasets.

4.1.1   PROPOSED APPROACH

In this subsection, we propose a novel graph-based transfer learning approach, which takes into consideration the human factor by modeling the task relatedness via both the shared users and keywords from both the domains.

4.1.1.1   Notation

Let $\mathscr{X}^S$ denote the set of examples from the source domain, i.e $\mathscr{X}^S = \{x_1^S, ..., x_m^S\} \subset \mathbb{R}^{ds}$, where m is the number of examples from the source domain, and ds the dimensionality of the feature space. Let $\mathscr{Y}^S$ denote the labels of these examples, i.e $\mathscr{Y}^S = \{y_1^S, ..., y_m^S\} \subset \{-1, 1\}^m$, where y_i^S is the class label of x_i^S, $1 \leq i \leq m$. Similarly for the target domain $\mathscr{X}^T$ denote the set of examples from the target domain, i.e $\mathscr{X}^T = \{x_1^T, ..., x_n^T\} \subset \mathbb{R}^{dt}$, where n is the number of examples from the target domain, and dt the dimensionality of the feature space. Let $\mathscr{Y}^T$ denote the labels of target domain examples, i.e $\mathscr{Y}^T = \{y_1^T, ..., y_{\varepsilon n}^T\} \subset \{-1, 1\}^{\varepsilon n}$, where y_i^T is the class label of x_i^T, $1 \leq i \leq \varepsilon n$. Let $d = ds \cup dt$ be the combined feature space for the source and target domains. For convenience we represent the features in the shared feature space of size d. Let $\mathscr{U}$ denote the set of users who posted the content of examples both in the source and target domains, i.e $\mathscr{U} = \{u_1, ..., u_u\} \subset [0, 1]^u$, where u is the number of unique users from the source and target domain. Among the target domain examples only the first εn are labeled, and $\varepsilon = 0$ corresponds to no labels from the target domain. Let $e = m + n$ be the total number of examples in source and target domain combined. Further the examples are split into labeled examples $e = m + \varepsilon n$ and unlabeled examples $eu = (1 - \varepsilon)n$. Our goal is to find a sentiment classification function $f_{eu} \rightarrow \{y_{\varepsilon n+1}^T, ..., y_n^T\}$ for all the unlabeled examples in the target domain $\mathcal{X}^T$ with a small error rate. Table 4.1 describes the notation for the proposed approach.

Table 4.1

Notation for the Proposed *U-Cross* Framework

Notation	Description
$\mathcal{X}^S$	Set of m examples from the source domain with d features.
$\mathcal{Y}^S$	Labels of m examples from the source domain.
$\mathcal{X}^T$	Set of n examples from the target domain with d features.
$\mathcal{Y}^T$	Labels of n examples from the target domain.
$\mathcal{U}$	Set of users from the source and target domains combined.
e, el, eu	# of examples, # of labeled examples and # of unlabeled examples respectively
$\mathbf{A}_{(3)}, \mathbf{D}_{(3)}, \mathbf{S}_{(3)}$	Affinity, Degree and Symmetric laplacian matrices

4.1.1.2   User-Example-Feature Tripartite Graph

The tripartite graph consists of three different types of nodes: users, examples and keyword features extracted from examples of both the domains. Let $G^{(3)} = \{V^{(3)}, E^{(3)}\}$ denote the undirected tripartite graph, where $V^{(3)}$ is the set of nodes in the graph, and $E^{(3)}$ is the set of weighted edges. Users are connected to examples in the source and target domain, i.e. there exists an edge between every example and the user who posted the example. Moreover, it is also possible to have a set of users who have examples only in the source domain or target domain but not in both. All the labeled and unlabeled example nodes are connected to corresponding feature nodes, i.e. there exists an edge between every labeled or unlabeled node to a feature node only if the feature has a positive weight associated with that example. The labeled and the unlabeled example nodes are not connected to each other. The edges between user nodes and examples have a weight $v_j \subset [0, 1]$. In the case of example and feature nodes, the edge weights can either be a real value or binary values. To explain this with regards to the sentiment classification task and real data, examples correspond to Amazon reviews, features represent the n-gram keywords of each review and user is the one who wrote the review. Figure 4.1 shows the example of a user-example-feature tripartite graph.

Given the tripartite graph $G^{(3)}$ we define the symmetric affinity matrix $\mathbf{A}^{(3)}$ of size $(u + e + d)$. The first u nodes correspond to the users, the next e nodes correspond to examples and the last d nodes represent keyword features extracted from examples. Considering m examples from the source domain and n examples from the target domain, the e examples consist of $el = m + \varepsilon n$ labeled examples followed by $eu = n - \varepsilon n$ unlabeled examples. The affinity matrix has the following structure:

$$\mathbf{A}^{(3)} = \begin{bmatrix} \mathbf{0}_{u \times u} & \mathbf{A}^{(1,2)}_{u \times e} & \mathbf{0}_{u \times d} \\ \mathbf{A}^{(2,1)}_{e \times u} & \mathbf{0}_{e \times e} & \mathbf{A}^{(2,3)}_{e \times d} \\ \mathbf{0}_{d \times u} & \mathbf{A}^{(3,2)}_{d \times e} & \mathbf{0}_{d \times d} \end{bmatrix}$$

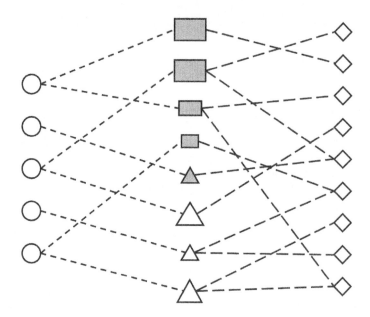

Figure 4.1 User-example-feature tripartite graph. Circles represent users; squares are the source domain examples; triangles are the target domain examples; diamonds represent the keywork features. Different sizes of squares represent the reweighted source domain examples. Filled squares and triangles represent the labeled examples.

where $\mathbf{0}_{a \times b}$ is an $a \times b$ zero matrix, $\mathbf{A}_{u \times e}$ is a non-zero user-example affinity matrix, $\mathbf{A}_{e \times d}$ is a non-zero example-keyword affinity matrix. $\mathbf{A}_{u \times e}$ and $\mathbf{A}_{e \times d}$ are the sub-matrices of the affinity matrix $\mathbf{A}^{(3)}$. The matrix $\mathbf{A}^{(3)}$ is a symmetric matrix such that $\mathbf{A}_{i,j} = \mathbf{A}_{j,i}$ where $\mathbf{A}_{i,j}$ is a submatrix of $\mathbf{A}^{(3)}$. We also define a diagonal matrix $\mathbf{D}^{(3)}$ of size $(u + e + d)$ with a diagonal element $\mathbf{D}_i^{(3)} = \sum_{j=1}^{u+e+d} \mathbf{A}_{i,j}^{(3)}, i = 1, ..., u + e + d$, where $\mathbf{A}_{i,j}^{(3)}$ denote the element in the ith row and j^{th} column of $\mathbf{A}^{(3)}$. The diagonal matrix has the following structure:

$$\mathbf{D}^{(3)} = \begin{bmatrix} \mathbf{D}_{u \times u}^{(1,1)} & \mathbf{0}_{u \times e} & \mathbf{0}_{u \times d} \\ \mathbf{0}_{e \times u} & \mathbf{D}_{e \times e}^{(2,2)} & \mathbf{0}_{e \times d} \\ \mathbf{0}_{d \times u} & \mathbf{0}_{d \times e} & \mathbf{D}_{d \times d}^{(3,3)} \end{bmatrix}$$

where $\mathbf{D}^{(1,1)}$, $\mathbf{D}^{(2,2)}$ and $\mathbf{D}^{(3,3)}$ are submatrices of diagonal matrix $\mathbf{D}^{(3)}$ which equals row sums of affinity submatrices $\mathbf{A}^{(1,2)}$, $(\mathbf{A}^{(1,2)})^T + \mathbf{A}^{(2,3)}$ and $(\mathbf{A}^{(2,3)})^T$ respectively. Finally we define $\mathbf{S}^{(3)} = (\mathbf{D}^{(3)})^{-1/2}\mathbf{A}^{(3)}(\mathbf{D}^{(3)})^{-1/2}$. Similar to $\mathbf{A}^{(3)}$, $\mathbf{S}^{(3)}$ is a symmetric matrix with non-negative elements $\mathbf{S}_{i,j}^{(3)}$ such that the sub matrices $\mathbf{S}^{(1,2)} = (\mathbf{S}^{(2,1)})^T$ and $\mathbf{S}^{(2,3)} = (\mathbf{S}^{(3,2)})^T$.

4.1.1.3 Objective Function

The goal of building a tripartite graph is to learn the sentiment classification function on unlabeled target domain data. We define four functions f_u, f_{el}, f_{eu} and f_d that take values on users, labeled examples from the source and target domains, unlabeled examples from the target domain and feature nodes respectively, and define f as: $f = [(f_u)^T, (f_{el})^T, (f_{eu})^T, (f_d)^T]^T$. We also define four column vectors y_u, y_{el}, y_{eu} and y_f of size u, el, eu and d respectively. We merge all the column vectors into a single column vector $y = [(y_u)^T, (y_{el})^T, (y_{eu})^T, (y_d)^T]^T$.

Example function f_{el} and column vector y_{el} are composed of the first $m + \varepsilon n$ values for labeled examples; similarly function f_{eu} and column vector y_{eu} are composed of $n - \varepsilon n$ values for unlabeled examples. Vectors y_u, y_{el}, y_{eu} and y_d represent the prior knowledge of users, labeled examples, unlabeled examples and features respectively. If we do not have any prior knowledge we set the vectors to zero. The vector y_{el} is set to sentiment labels $\{-1, 1\}$ corresponding to the labeled examples.

In regular supervised learning problems the training data and test data are usually from the same distribution, but in a situation when training data and test data are from different distributions, it is called covariate shift. In transfer learning tasks, distribution of data in the source domain varies with distribution of data in the target domain. In such scenarios reweighting training data $w(x) = p_{test}(x)/p_{train}(x)$ to fit test data distribution often resulted in increased classification performance [Bickel et al., 2009, Huang et al., 2006, Sugiyama et al., 2008]. We used the reweighting technique as suggested in Sugiyama et al. [2008] to reweight the source domain examples based on entire set of examples from the target domain.

We propose to minimize the following objective function with respect to f.

$$Q_1(f) = \frac{1}{2} \sum_i^e w_i \sum_j^u v_j \mathbf{A}_{i,j}^{(3)} \left(\frac{f_i}{\sqrt{\mathbf{D}_i^{(3)}}} - \frac{f_j}{\sqrt{\mathbf{D}_j^{(3)}}} \right)^2$$

$$+ \frac{1}{2} \sum_i^e w_i \sum_j^d \mathbf{A}_{i,j}^{(3)} \left(\frac{f_i}{\sqrt{\mathbf{D}_i^{(3)}}} - \frac{f_j}{\sqrt{\mathbf{D}_j^{(3)}}} \right)^2$$

$$+ \mu \sum_k^{u+e+d} (f_k - y_k)^2$$

$$= f^T (I_{(u+e+d)\times(u+e+d)} - \mathbf{S}^{(3)})f + \mu ||f - y||^2$$

where w_i is the example reweighting parameter to reduce the covariate shift between the source and target domain examples, $w_i = 1$ for the target domain examples, v_j is the user soft-score weight to ensure user consistency across the source and target domains, μ is a small positive parameter and I is the identity matrix. The objective function has three terms. The first and second terms in the equation measures the label smoothness of the function f w.r.t users with labeled examples and keywords with labeled examples respectively. The second term represents the consistency of the function f with label information and prior knowledge.

4.1.1.4   User Soft-Score Weights

Our proposed approach utilizes user behavior from labeled examples in computing the sentiment of the posts from the target domain. It is very likely that the sentiment labeling behavior of a user might vary across the source and target domains. For example, it is possible that a certain user has more positive reviews in the source domain and more negative reviews in the target domain. Such users degrade the performance of the classifier due to inconsistency in user behavior across the source and target domains. In extreme cases such inconsistency might lead to negative transfer learning.

In our approach we handle this issue by assigning non-negative soft-weights $v_{cu} \in [0,1]$ to the set of common users $cu \in \mathcal{U}_c$ and $\mathcal{U}_c \subseteq \mathcal{U}$ from the source and target domains. We use the labeled examples from the source and target domains along with their keywords and sentiment labels to assign a soft-score to each shared user. The user soft-score weight calculation mechanism for each shared user across domains is as follows:

$$v_{cu} = \sum_i^{el^S} \sum_j^{el^T} sim(\mathbf{x}_i, \mathbf{x}_j) * y_i * y_j \tag{4.1}$$

where el^S and el^T represent the set of labeled examples for the user u in the source and target domains respectively, $\mathbf{x}_i \in \mathcal{X}^S$ and $\mathbf{x}_j \in \mathcal{X}^T$ represent the feature vectors for the examples in the source and target domains, $sim(\mathbf{x}_i, \mathbf{x}_j)$ is the cosine similarity between the feature vectors $\mathbf{x}_i$ and $\mathbf{x}_j$, finally, y_i and y_j are the corresponding sentiment labels for the examples i and j. In order to avoid negative transfer due to inconsistent user behavior across domains, the approach assigns smaller weights to inconsistent users. From eq (4.1), the more consistent users have a positive value and more inconsistent users have a negative value. As the edge weights are always positive, we scale the user weights v_{cu} from $[-1,1]$ to $[0,1]$.

4.1.1.5   U-Cross Algorithm

To minimize Q_1, we first set $f_{el} = y_{el}$, which requires the outputs of the classification function to be consistent with the known labels in the source and target domains, and then solve for f_{eu}, f_u, and f_d from the following lemma.

Lemma 4.1

If $f_{el} = y_{el}$, Q_1 is minimized at:

$$f_{eu}^* = [\alpha(\mathbf{S}_{eu}^{(1,2)})^T \mathscr{P} + \alpha(\mathbf{S}_{eu}^{(1,2)})^T \mathscr{P} + R]$$
$$[I - \alpha^2(\mathbf{S}_{eu}^{(1,2)})^T \mathbf{S}_{eu}^{(1,2)} - \alpha^2 \mathbf{S}_{eu}^{(2,3)}(\mathbf{S}_{eu}^{(2,3)})^T]^{-1} \tag{4.2}$$

$$f_u^* = \mathscr{P} + \alpha \mathbf{S}_{el}^{(1,2)} f_{eu}^* \tag{4.3}$$

$$f_d^* = \mathcal{Q} + \alpha (\mathbf{S}_{el}^{(2,3)})^T f_{eu}^* \tag{4.4}$$

where $\alpha = \frac{1}{1+\mu}$, $\mathcal{P} = \alpha \mathbf{S}_{el}^{(1,2)} y_{el} + (1-\alpha) y_u$, $\mathcal{Q} = \alpha (\mathbf{S}_{el}^{(2,3)})^T y_{el} + (1-\alpha) y_d$ and $\mathcal{R} = (1-\alpha) y_{eu}$ ∎

Proof. After setting $f_{el} = y_{el}$ in Q_1, we get

$$\begin{aligned}
Q_1 &= f_u^T f_u + y_{el}^T y_{el} + f_{eu}^T f_{eu} + f_d^T f_d \\
&\quad - 2f_3^T S^{(2,3)} f_d - 2f_u^T S^{(1,2)} y_{el} - 2f_u^T S^{(1,2)} f_{eu} \\
&\quad - 2y_{el}^T S^{(2,3)} f_d + \mu \|f_u - y_u\|^2 + \mu \|f_{eu} - y_{eu}\|^2 \\
&\quad + \mu \|f_d - y_d\|^2
\end{aligned} \tag{4.5}$$

Minimizing by partial differentiation w.r.t f_u, f_{eu} and f_d equating the terms to zero, we can find the optimal values f_u^*, f_{eu}^* and f_d^*.

$$\frac{\partial Q_1}{f_u} = 2f_u - 2S^{(1,2)} y_{el} - 2S^{(1,2)} f_{eu} - 2\mu y_u$$

$$\frac{\partial Q_1}{f_d} = 2f_d - 2S^{(2,3)} y_{el} - 2S^{(2,3)} f_{eu}) - 2\mu y_d$$

$$\frac{\partial Q_1}{f_{eu}} = 2f_{eu} - 2(S^{(1,2)})^T f_u - 2(S^{(3,2)})^T f_d$$
$$- 2\mu y_{eu}$$

Equating the above equations to zero, and substituting the parameter $\alpha = \frac{1}{1+\mu}$, we get equations 3, 4 and 5:

$$f_u = \alpha S_{el}^{(1,2)} y_{el} + \alpha S_{eu}^{(1,2)} f_{eu} + (1-\alpha) y_u \tag{4.6}$$

$$f_d = \alpha (S_{el}^{(2,3)})^T y_{el} + \alpha (S_{eu}^{(2,3)})^T f_{eu} + (1-\alpha) y_d \tag{4.7}$$

$$f_{eu} = \alpha (S_{eu}^{(1,2)})^T f_u + \alpha S_{eu}^{(2,3)} f_d + (1-\alpha) y_{eu} \tag{4.8}$$

Solving equations (4.6), (4.7) and (4.8) leads to the optimal values f_u^*, f_{eu}^* and f_d^*. □

From equations (4.2), (4.3) and (4.4) computing f_u^*, f_{eu}^* and f_d^* requires solving matrix inversions which is a computationally intensive operation given the large size of the unlabeled examples and keyword features. To address this issue we consider the following iteration steps obtained after minimizing Q_1 to compute the optimal solution.

$$f_{eu}(t+1) = (1-\alpha) y_{eu} - \alpha ((S_{eu}^{(1,2)})^T f_u(t) + S_{eu}^{(2,3)} f_d(t)) \tag{4.9}$$

$$f_u(t+1) = (1-\alpha) y_u - \alpha (S_{el}^{(1,2)} y_L + S_{eu}^{(1,2)} f_{eu}(t)) \tag{4.10}$$

$$f_d(t+1) = (1-\alpha) y_d - \alpha ((S_{el}^{(2,3)})^T y_{el} + (S_{eu}^{(2,3)}))^T f_{eu}(t)) \tag{4.11}$$

where t is the number of iterations. The following theorem guarantees the convergence of these iteration steps:

Theorem 4.1

When t goes to infinity, $f_{eu}(t)$ converges to f_{eu}^*, $f_u(t)$ converges to f_u^* and $f_d(t)$ converges to f_d^*. ∎

Proof. Substituting equations (4.10) and (4.11) into (4.9),

$$f_{eu}(t) = \alpha^2[(S_{eu}^{(1,2)})^T S_{eu}^{(1,2)} + S_{eu}^{(2,3)}(S_{eu}^{(2,3)})^T]f_{eu}(t-2)$$
$$+ R + \alpha(S_{eu}^{(1,2)})^T P + \alpha S_{eu}^{(2,3)} Q$$
$$= \alpha^2 \mathscr{G} f_{eu}(t-2) + \mathscr{H}$$
$$= (\alpha^2 \mathscr{G})^{\frac{t}{2}} f_{eu}(0) + \left(\sum_{i=0}^{\frac{t}{2}-1} (\alpha^2 \mathscr{G})^i\right)\mathscr{H}$$

where $\mathscr{G} = [(S_{eu}^{(1,2)})^T S_{eu}^{(1,2)} + S_{eu}^{(2,3)}(S_{eu}^{(2,3)})^T]$ and $\mathscr{H} = R + \alpha(S_{eu}^{(1,2)})^T P + \alpha S_{eu}^{(2,3)} Q$. Since $\mu > 0$ and $\alpha = \frac{1}{1+\mu}$, $0 < \alpha < 1$. As given in Lemma 2 of the paper He et al. [2009], the eigenvalues of $\mathscr{G}$ are in [-1,1]. Therefore we have,

$$\lim_{t \to \infty} (\alpha^2 \mathscr{G})^{\frac{t}{2}} f_{eu}(0) = 0$$

$$\lim_{t \to \infty} \left(\sum_{i=0}^{\frac{t}{2}-1} (\alpha^2 \mathscr{G})^i\right) = (I - \alpha^2 \mathscr{G})^{-1}$$

Hence, if t is an even number,

$$\lim_{t \to \infty} f_{eu}(t) = f_{eu}(t)^*$$

□

Based on the above discussion, we present the *U-Cross* algorithm in Algorithm 1. Our algorithm *U-Cross* takes as input a set of m labeled examples as an example-keyword sparse binary matrix from the source domain, set of n examples as an example-keyword sparse binary matrix from the target domain among which a small subset $n\varepsilon$ are labeled examples and the set of users $\mathscr{U}$ who authored the examples from the source and target domains. The algorithm outputs the labels of all the unlabeled examples from the target domain.

As an initial data processing step, we construct the affinity matrix $\mathbf{A}^{(3)}$ from the user-example and example-keyword affinity matrices, and then compute the degree

---

Algorithm 1: *U-Cross* Algorithm

Input: Set of m labeled examples from source domain $\mathscr{X}^S$ and their labels $\mathscr{Y}^S$; set of n examples from target domain $\mathscr{X}^T$ and labels for first εn examples $\mathscr{Y}^T$; users who authored the examples $\mathscr{U}$; the number of iterations t.

Output: Labels of all unlabeled examples in $\mathscr{X}^T$

1 Calculate the soft-score weights v_u for all the shared users according to eq (4.1). Set weights of user-user adjacency matrix $\mathbf{A}_u$ from v_u.

2 Set labeling function f_{el} to given labels y_{el}, $f_{el} = y_{el}$; Set initial user information y_u, unlabeled values y_{eu} and feature values y_d to zero if their prior values are not available. Initialize the corresponding functions $f_u(0)$, $f_{eu}(0)$ and $f_d(0)$ to y_u, y_{eu} and y_d respectively.

3 **for** $i \leftarrow 1$ **to** t **do**

4 $\quad$ Calculate $f_u(i)$ and $f_d(i)$ according to eq (4.10) and eq (4.11).

5 $\quad$ Calculate $f_{eu}(i)$ according to eq (4.9) and using the functions $f_u(i)$ and $f_d(i)$ calculated in previous step.

6 **end**

7 **for** $i \leftarrow (\varepsilon n + 1)$ **to** n **do**

8 $\quad$ If $f_{eu}(t)$ at $x_i^T > 0$ then set $y_i^T = 1$ else set $y_i^T = -1$

9 **end**

---

matrix $\mathbf{D}^{(3)}$ and normalized symmetric matrix $\mathbf{S}^{(3)}$. As a preprocessing step, we calculate the covariate shift parameter weights w_i as discussed in Subsection 4.1.1 to reweight all the source domain examples. In Step 1, we calculate the soft-score weights for all the shared users across the source and target domains to ensure consistency in sentiment labeling behaviors. As the only known prior values are the labels from the source domains, we initialize the function for labeled examples f_{el} to the known set of labels y_{el} and initialize the rest of the prior values and corresponding functions to 0. In Step 2, we learn the functions for users, unlabeled examples and keywords by label propagation using the gradient method over t iterations. The functions are updated using eq (4.10), eq (4.11) and eq (4.9). Finally, in the last step the sign of the function value for each unlabeled example is set as the sentiment label.

Based on the notation in Section 4.1.1.1, the following lemma shows the computational complexity of *U-Cross*:

Lemma 4.2

The computational complexity of the *U-Cross* is given by $\mathscr{O}\big(t(n+m)(u+d) + (p_{max})^2 * (d_{max})^2 * u\big)$, where t is the number of *U-Cross* iterations, p_{max} is the maximum number of posts generated by a user, and d_{max} is the maximum number of keywords in a post. ∎

Proof. The proposed *U-Cross* algorithm has two main steps. The first step is to calculate the soft-score weights for all the shared users which takes about $\mathscr{O}((p_{max})^2 * (d_{max})^2 * u)$ time. In general, the cosine similarity function takes $\mathscr{O}((d_{max})^2)$ in the worst case. For all the data sets we choose, the number of posts for each user is no more than p_{max}. So calculating the pairwise similarities between posts of a user will result in less than $(p_{max})^2$ cosine similarity calculations. The second step is to find the function on unlabeled target domain data through a set of iterative updates. The time complexity for the updates with a set of matrix multiplications at its core is $\mathscr{O}\left(t(n+m)(u+d)\right)$. The last step is to assign the class labels to the target domain unlabeled examples based on the sign of function f_{eu} with a time complexity of $\mathscr{O}(n)$. □

From this lemma, we can see that *U-Cross* scales linearly with respect to the problem size (e.g., the number of examples in the source domain and the target domain, the size of the combined vocabulary space). Therefore, it can be naturally applied to large datasets.

4.1.2   CASE STUDY

In this subsection we discuss how an existing method named TRITER [He et al., 2009] can be seen as a special case of *U-Cross*. TRITER uses both the keywords-labeled-unlabeled examples tripartite graph and a labeled-unlabeled examples bipartite graph to model the relationship between the source and the target domains, using high weights for examples from the target domain. However, in scenarios where a target domain example is mapped to both positive and negative examples from the source domain, the inclusion of the bipartite graph could even harm the performance. Therefore, we ignore the bipartite graph (i.e., setting the corresponding weight to 0), and use a reweighting scheme to connect examples from the source domain and the target domain. More specifically, using the same notation as in the previous section, the objective function of TRITER can be written as follows.

$$
\begin{aligned}
Q_2(f) = {} & \frac{1}{2}\sum_i^e w_i \sum_j^d A_{i,j}^{(3)} \left(\frac{f_i}{\sqrt{D_i^{(3)}}} - \frac{f_j}{\sqrt{D_j^{(3)}}} \right)^2 \\
& + \mu \sum_i^{e+d} (f_k - y_k)^2 - \gamma \sum_l^e (f_l^T f_{u(l)})^2 \\
= {} & f^T (I_{(e+d)\times(e+d)} - S^{(3)})f + \mu\|f - y\|^2 \\
& + \beta\|f_U\|^2 - \gamma(\|f_{el}^T U_{el}^T f_U\|^2 + \|f_{eu}^T U_{eu}^T f_{eu}\|^2)
\end{aligned}
$$

where μ, β and γ are positive parameters, w_i is the instance weight for labeled and unlabeled nodes and I is an identity matrix. U_{el} and U_{eu} are matrices of size $u \times m$ and $u \times n$ respectively that map users to labeled and unlabeled examples. Matrices U_{el} and U_{eu} can be compared to $A_{el}^{(1,2)}$ and $A_{eu}^{(1,2)}$ matrices in the tripartite graph mentioned in the previous section. The extension includes adding a regularizer on user behavior function and also on the user-example interaction. The last equation in minimization function Q_2 captures the interaction between users and different labeled and unlabeled examples in the graph which needs to be maximized.

Comparing all the terms in Q_1 and Q_2, we can see that both equations are similar. By setting $\beta = 1$ and $\gamma = 2$, it is possible to rewrite equation Q_2 in terms of Q_1 with minimal difference. The major difference between *U-Cross* and TRITER is that TRITER does not model user behavior. From the objective function Q_2, it can be seen that TRITER is a special case of *U-Cross* without user behavior. Therefore, *U-Cross* is expected to perform better that TRITER since it explicitly models the human factor.

4.1.3 RESULTS

In this subsection we report the experimental results. We first introduce three real-world cross-domain sentiment datasets related to product reviews. Then we compare different user soft-weight scoring approaches. Finally, we compare the *U-Cross* with other state-of-the-art methods to demonstrate its effectiveness.

4.1.3.1 Data Sets

To compare our transfer learning approach *U-Cross* we perform experiments on three different real-world datasets. User reviews from three different product review websites are used for the sentiment classification task. Table 4.2 describes the dataset statistics. The dataset details are as follows:

1. Amazon product reviews[2]: The dataset is a part of the Stanford Network Analysis Project [McAuley and Leskovec, 2013a] and includes Amazon product reviews from 28 different product categories. For experimental evaluation, we created six different datasets with varying common user frequency from *office products, software, toy games, video games, electronics, videos, kitchen, movies* and *music* product categories.
2. Yelp reviews[3]: The data set is from the Yelp Data set challenge and includes user reviews from *restaurants* and *shopping* domains.
3. CIAO dataset[4]: The dataset is crawled from the CIAO website and consists of consumer reviews of *books* and *beauty* products.

Several preprocessing steps were taken before experiments. Words were converted to lower cases and then stemmed to terms. All the stop words, punctuation and symbols were removed. Binary feature vector as a bag of words on n-grams $n = \{1, 2, 3\}$ was extracted for each review. Also, we dropped those users with less than three reviews and more than a hundred reviews in the source and target domains

---

[2]http://www.amazon.com

[3]http://www.yelp.com/dataset_challenge

[4]http://www.ciao.co.uk

Table 4.2

Dataset Statistics

	Source domain	Target domain	Examples	Examples	Unique Users	Source Users	Target Users	Common Users	Features
	office products	software	2225	3968	1362	588	832	58	9069
	toy games	video games	5760	16700	3937	1153	3127	343	26982
Amazon	electronics	amazon video	41340	72976	21102	9632	12770	1300	144669
	electronics	kitchen	44518	56296	15126	9279	8952	3105	118418
	amazon videos	music	73448	478414	78842	11538	74149	6845	1704032
	amazon videos	movies	73177	533418	82665	17799	79677	14811	1908543
Yelp	restaurants	shopping	5502	19338	622	510	345	310	95094
CIAO	beauty	books	3524	6734	339	217	195	73	64429

to ensure consistent and unbiased user contribution. Features with document (product reviews) frequency less than 10 were also dropped. Table 4.2 reports the size of the feature vector on the entire vocabulary space for the source and target domains combined. Moreover as explained in Section 4.1.1.5, the source domain examples were reweighted to reduce the covariate shift across the domains.

4.1.3.2 User Selection

As prior research demonstrated that by using user information along with linguistic features improved the performance of sentiment classifiers, we employ a robust user selection approach proposed in Section 4.1.1.4 to assign soft-score weights to all the common set of users $\mathcal{U}_c$ in the source and the target domains. In order to avoid negative transfer due to inconsistent user behavior across domains, the approach assigns larger weights to more consistent users with similar product labeling behavior across domains and smaller weights to inconsistent users.

Let $\mathcal{U}^S$ and $\mathcal{U}^T$ be the set of unique users in the source and target domains respectively. To evaluate the effectiveness of our user selection approach, we consider the following variations of the soft-score user weights:

1. **USW1**: The baseline approach that assigns unit weights to all the users $\mathcal{U}$.

$$v_u = 1, \forall u \in \{\mathcal{U}^S \cup \mathcal{U}^T\}$$

2. **USW2**: Set all the user weights for shared users across domains as per the proposed approach and the rest to 0.

$$v_u = \begin{cases} v_{cu}, & \forall u \in \mathcal{U}_c \\ 0, & \forall u \in \{\mathcal{U}^S \cup \mathcal{U}^T\} \backslash \mathcal{U}_c \end{cases}$$

3. **USW3**: Set all the user weights for shared users across domains as per the proposed approach and the non-shared users in the target domain to 1.

$$v_u = \begin{cases} v_{cu}, & \forall u \in \mathcal{U}_c \\ 0, & \forall u \in \mathcal{U}^S \backslash \mathcal{U}_c \\ 1, & \forall u \in \mathcal{U}^T \backslash \mathcal{U}_c \end{cases}$$

Figure 4.2 compares the performance of different user soft-score weighting approaches USW1, USW2 and USW3 on the Amazon reviews data set (*electronics → amazon videos*). It can be observed that USW3 performs the best compared to USW1 and USW2. In congruence with previous findings, leveraging the knowledge from associations between users and examples USW3 performed better compared to not using the associations between users and examples USW2. The approach with equal weights to all the users (consistent and inconsistent) performed the worst because of the negative transfer effect often associated with transfer learning. In all the following experiments, unless specified otherwise, the user soft-score weights in *U-Cross* refers to USW3.

4.1.3.3 Empirical Analysis

To show the robustness of the *U-Cross* approach we run the experiments by resampling the labeled examples from the target domain. We report the results with confidence scores from 20 runs. The target domain examples are carefully chosen to minimize the class- and user-bias. By class bias we choose target domain examples with fairly equal proportions of positive and negative class labels. Also we ensure that the chosen labeled target examples maximize the set of common users across the source and target domains. We compare our *U-Cross* approach with other state-of-the-art methods and report the test error on the unlabeled examples from the target domain. The methods to be compared include: SCL [Blitzer et al., 2011], where for each data

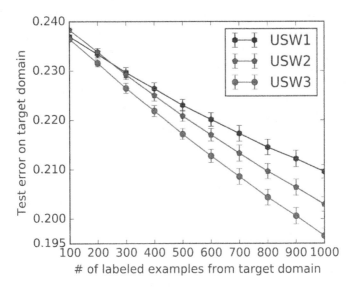

Figure 4.2 Comparison of user selection approaches USW1, USW2 and USW3. The x-axis represents # of labeled examples from the target domain and the y-axis represents test error on the target domain data.

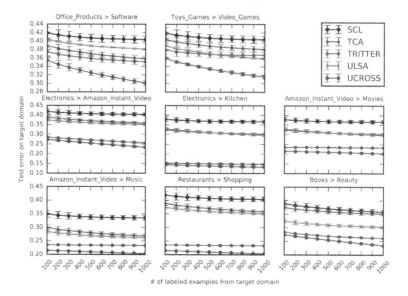

Figure 4.3 Performance evaluation on various datasets. In each subfigure, the title represents the source and target domains, the x-axis represents # of labeled examples from the target domain and the y-axis represents test error with error bars on the target domain over 20 runs. In each run the labeled data from target domain is resampled.

Table 4.3
Performance Comparison of *U-Cross* with Other Methods

Source domain	Target domain	SCL	TCA	TRITER	ULSA	U-Cross
Office Products	Software	0.414 ± 0.007	0.358 ± 0.005	0.349 ± 0.006	0.381 ± 0.001	**0.301 ± 0.005**
Toy Games	Video Games	0.430 ± 0.006	0.381 ± 0.005	0.367 ± 0.006	0.358 ± 0.001	**0.315 ± 0.004**
Electronics	Amazon Videos	0.424 ± 0.005	0.351 ± 0.003	0.255 ± 0.001	0.351 ± 0.003	**0.234 ± 0.001**
Electronics	Kitchen	0.367 ± 0.003	0.298 ± 0.004	0.147 ± 0.001	0.302 ± 0.004	**0.132 ± 0.001**
Amazon Videos	Music	0.367 ± 0.004	0.298 ± 0.003	0.233 ± 0.001	0.302 ± 0.002	**0.203 ± 0.001**
Amazon Videos	Movies	0.335 ± 0.003	0.268 ± 0.004	0.221 ± 0.001	0.261 ± 0.003	**0.196 ± 0.001**
Restaurants	Shopping	0.404 ± 0.006	0.358 ± 0.004	0.228 ± 0.001	0.351 ± 0.004	**0.201 ± 0.001**
Beauty	Books	0.358 ± 0.005	0.351 ± 0.004	0.261 ± 0.001	0.302 ± 0.003	**0.234 ± 0.001**

set, 2000 pivot features are selected from the source and target domains; TCA [Li et al., 2012], which utilizes both the shared and the mapped domain-specific topics to span a new shared feature space for knowledge transfer; TRITER which leverages labeled-unlabeled-keywords to propagate sentiment information from labeled examples to unlabeled examples; and ULSA [Tan et al., 2011], which performs user-level sentiment analysis incorporating social networks with user-user relationship parameter $\lambda_k = 0$.

The parameter selection for *U-Cross* is performed through 10-fold cross validation on labeled examples from the source and target domains on different datasets. Labeled examples from the target domain are randomly sampled over 20 runs to ensure robust parameter selection. A total of 1000 examples are sampled from the target domain for parameter selection. From the empirical results, setting the regularization parameter to $\alpha = 0.1$ resulted in best performance. So, we have set the regularization parameter of *U-Cross* to $\alpha = 0.1$ for comparison with other state-of-the-art methods.

The experimental results are summarized in Table 4.3. Figure 4.3 compares the performance of *U-Cross* on different datasets from Table 4.2. In each figure, the x-axis represents number of labeled examples from the target domain and the y-axis represents test error with error bars on the target domain over 20 runs. First, our *U-Cross* approach outperforms all other methods in terms of test error on unlabeled examples from the target domain in all the datasets. This validates the effectiveness of leveraging user information for cross-domain sentiment classification of user reviews. Second, the variation is significant in datasets with the large user network which shows that user behavior plays a significant role in large scale sentiment classification tasks.

4.2 SIMILAR ACTOR RECOMMENDATION

Cross-network link prediction has immense applications in the current social media driven world. For example, Tang et al. [2015a] showed that document level sentiment classification can be improved by associating users across networks; through cross-network link prediction, better expert matching for collaborations in academia is made possible by integrating the author collaboration networks with professional networks like ResearchGate and LinkedIn [Zhang et al., 2015]; a good music recommender system can be built by studying music interests of users across Twitter, Last.fm and Facebook [Su et al., 2013].

Cross-network link prediction is often challenging due to the sparsity of the cross-network data. In most cases, the links across social networks are either extremely sparse or non-existent making it difficult to learn associations from one network to the other. We address this issue by leveraging the shared features across the social networks. We assume that there exists a shared feature space across the social network based on which associations can be learned. In the case of diabetes forum posts, the users post about their diabetes information like symptoms, self-care habits and health care measures which have common keywords across the networks. The goal of the algorithm is to first learn the latent features of the keywords and then learn the associations between the members of the social network through keyword latent features of the shared feature space. The goal of the algorithm is to learn associations between the members of the social network through associations learned from shared features. It can be realized that the performance of the cross-network link prediction algorithm depends on the quality of the shared feature space.

In this section, we propose the CrossNet framework to predict links between similar actors across networks. We organize social relations as a user-user bipartite graph and user-generated keywords as a user-keyword bipartite graph for each social network. We perform co-clustering on the user-keywords graph to learn the representation of users and keywords in a latent feature space. We propose a constrained non-negative matrix tri-factorization (NMTF) approach with a graph regularizer to integrate such graphs from multiple social networks. We predict the links between the similar actors across the networks through the respective user latent features learned from each network and user-user associations in each network. We employ Random walks with restarts (RWR) Sun et al. [2005b] to learn the cross-network user-user associations. Notice that this problem is different from cross-domain link prediction, where given social networks from different domains, the task is to leverage the knowledge from one social network to predict the missing links in another social network. Also this problem is different from author link prediction, which involves identifying user's accounts in multiple social networks.

The main contributions of this section are as follows:

1. CrossNet – a novel non-negative matrix tri-factorization based approach to co-cluster users and keywords from multiple social networks simultaneously.
2. As the proposed problem is non-convex, we propose multiplicative updates to efficiently compute the user and keyword latent features.
3. We demonstrate the effectiveness of CrossNet on the real world academic networks data set. Also we perform a case study on the diabetes social networks data set.

Nowadays, online social networks have become an important portal for patients with major diseases, such as diabetes mellitus, to connect with physicians as well as other patients. Compared with the generic social networks such as Twitter and Facebook, the disease-specific social networks (e.g., TuDiabetes[5] and DiabetesSisters[6]) have a greater concentration of patients with similar conditions, and the patients expect to obtain additional resources from these social networks. However, when it comes to using these social networks, it is often the case that a patient would stick to a single social network, and rarely look at the other social networks, thus limiting their access to the online resources, especially the patients with similar questions and concerns from the other social networks. Motivated by this application, in this section, we focus on cross network link recommendation, which aims to identify similar actors across multiple heterogeneous social networks. In this way, we will be able to form support groups consisting of patients from multiple disease-specific networks, all sharing the same questions and concerns.

The problem setting studied in this section is similar and yet significantly different from existing work on cross network link prediction. In particular, existing work

---

[5]http://www.tudiabetes.org/

[6]https://diabetessisters.org/

either links different accounts belonging to the same user across multiple social networks Zhang et al. [2015], or links users with complementary expertise or interest Tang et al. [2012]. In contrast, we aim to find *similar* users using different social networks, which enables them to exchange important information regarding their shared questions or concerns.

Based on the observation that different disease-specific social networks tend to share the same topics as well as the interests of user groups in certain topics, we propose to jointly decompose the user-keyword matrices from these social networks, while requiring them to share the same topics and user group-topic association matrices. To be specific, we form a generic optimization framework, and instantiate it with variations of the constraints. Then we propose an iterative optimization algorithm and analyze its performance from multiple perspectives. Finally, we test the performance of this algorithm on various real-world data sets, which outperforms state-of-the-art techniques.

4.2.1 PROBLEM DEFINITION

In this section, we formally introduce the cross network link recommendation problem, followed by the proposed generic optimization framework and its instantiations. Then we present the iterative optimization algorithm as well as its performance analysis.

4.2.1.1 Notation and Problem Definition

Suppose that we have K disease-specific social networks: $\mathcal{G}_k = \langle V_k^U, E_k^U \rangle$, $k = 1, \ldots, K$, where V_k^U is the set of user nodes $|V_k^U| = m_k$ and $E_k^U \subseteq V_k^U \times V_k^U$ is the set of edges representing the connection between users in the same social network. Self-connections and multiple links between two user nodes are not allowed. Let $\mathbf{A}_k \subset \{0,1\}^{m_k \times m_k}$ denote the user-user adjacency matrix for the kth social network $k = 1, \ldots, K$, where the edge weight is set to 1 if there is a connection between two users. Notice that we focus on the more challenging case where: (1) there are no shared user nodes across the social networks, i.e., $V_i^U \cap V_j^U = \emptyset, i \neq j \; \forall \; i, j = 1, \ldots, K$, and (2) there are no cross network links available between the users in different social networks. The goal of cross network link recommendation is to identify similar actors across multiple social networks. This is different from existing work on cross network link prediction which focuses on linking different accounts of the same user, or finding users with complementary expertise or interest.

Let $\mathcal{G}_k' = \langle V_k^U, V_k^W, E_k^{UW} \rangle$ denote the undirected user-keyword bipartite graph for the kth social network, where V_k^W is the set of keyword nodes $|V_k^W| = n_k$ and $E_k^{UW} \subseteq V_k^U \times V_k^W$ is the set of edges connecting the user nodes and the keyword nodes. Let $\mathbf{X}_k \subset \mathbb{R}^{m_k \times n_k}$ be the user-keyword adjacency matrix constructed from the bipartite graph $\mathcal{G}_k'$, $k = 1, \ldots, K$. Let d be the size of the vocabulary for all the social networks combined, i.e., $|V_1^W \cup V_2^W \cup .. \cup V_K^W| = d$.

Figure 4.4 illustrates the cross network link recommendation problem with two social networks $K = 2$. Figure 4.4(A) shows the user-user connection graphs $\mathcal{G}_1$

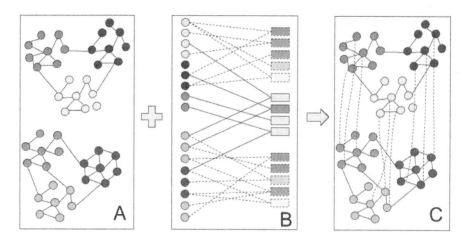

Figure 4.4   Cross network link prediction problem: A) two social networks with user nodes represented by circles and user-user associations represented by edges joining two nodes, different colors represent different user groups; B) user-keyword bipartite graph, circles represent users from different social networks, squares represent keywords from vocabulary space for different social networks. Dotted lines link the users to unique keywords in a social network and solid lines link users to shared keywords; C) dotted lines represent the recommended links between similar actors across social networks.

and $\mathscr{G}_2$. Figure 4.4(B) represents the user-keyword bipartite graphs $\mathscr{G}_1'$ and $\mathscr{G}_2'$. Figure 4.4(C) represents the problem of cross network link recommendation that recommends links between user nodes from different social networks $\mathscr{G}_1$ and $\mathscr{G}_2$.

Problem 1. *Cross network link prediction across multiple social networks.*

Input: The input to the problem is a set of user-user adjacency matrices $\{\mathbf{A}_1, \mathbf{A}_2, \ldots, \mathbf{A}_K\}$ constructed from user relationship graphs $\mathscr{G}_k, k = 1, \ldots, K$ and a set of user-keyword adjacency matrices $\{\mathbf{X}_1, \mathbf{X}_2, \ldots, \mathbf{X}_K\}$ constructed from user-keyword bipartite graphs $\mathscr{G}_k', k = 1, \ldots, K$.
Output: A set of cross network links $E^U \subseteq V_i^U \times V_j^U$ connecting similar user nodes V_i^U from the social network $\mathscr{G}_i$ to user nodes V_j^U from the social network $\mathscr{G}_j$, where $i \neq j$ and $i, j = 1, \ldots, K$.

4.2.2   PROPOSED APPROACH

4.2.2.1   Matrix Factorization for Cross Network Link Recommendation

In order to identify the similar actors across multiple disease-specific social networks, we propose to perform co-clustering on user-keyword graphs to learn the representation of users and keywords in a latent feature space, and then recommend the links between similar actors across the networks through the respective user latent features learned from each network. To be specific, we propose a constrained non-negative matrix tri-factorization (NMTF) approach with a graph regularizer obtained from the user-user adjacency matrices.

We begin by considering existing NMTF approaches and later introduce our approach for link recommendation. NMTF as shown in eq (4.12) involves decomposing a matrix $\mathbf{X} \subset \mathbb{R}^{m \times n}$, into three non-negative latent factor matrices $\mathbf{F} \subset \mathbb{R}_+^{m \times p}$, $\mathbf{S} \subset \mathbb{R}_+^{p \times o}$ and $\mathbf{G} \subset \mathbb{R}_+^{n \times o}$ that can best approximate $\mathbf{X}$. For example, in the context of social network analysis, given the user-keyword matrix for a social network, NMTF co-clusters users and keywords into p user groups and o keyword groups.

$$\mathbf{X} = \mathbf{FSG}^T \tag{4.12}$$

Cai et al. [2011] proposed a co-clustering method called Graph based non-negative matrix factorization (GNMF) that adds a graph regularizer to NMF imposing manifold assumptions. The factors for multiple social networks can be computed individually through K subproblems as follows:

$$\min \quad \left\| \mathbf{X}_k - \mathbf{F}_k \mathbf{G}_k^T \right\|_F^2 + \alpha_k \mathbf{tr}\left(\mathbf{F}_k^T \mathbf{L}_k \mathbf{F}_k \right)$$
$$\text{s.t.} \quad \mathbf{F}_k \geq 0, \ \mathbf{G}_k \geq 0, \ k = 1, \ldots, K \tag{4.13}$$

where $\mathbf{tr}(.)$ is the trace of the matrix, $\mathbf{L}_k = \mathbf{D}_k - \mathbf{A}_k$ is the graph Laplacian of user-user adjacency matrix $\mathbf{A}_k$, $\mathbf{D}_k = \sum_j \mathbf{A}_k^{ij}$ is the degree matrix, α_k is the regularization parameter on the user groups and $||.||_F^2$ is the Frobenius norm. The first term in the objective function minimizes the reconstruction error and the second term is a manifold regularizer on user-user relations which incorporates the geometric information of the data. If two users are closely connected to each other, they belong to the same group.

Gu et al. [2011] and Huang et al. [2014] showed that when regularization parameter α_k is set to a large value GNMF ends up in a trivial solution, associating all the users to one group. Also GNMF is prone to scale transfer problems, when the parameters in the objective function multiplied by any scalar ($\gamma > 1$) results in a solution which is different from the optimal solution. To fix these two issues, Gu et al. [2011] proposed a graph based NMTF approach (IGNMTF), with three factors and orthogonal constraints to allow more degrees of freedom between user and keyword latent factors. Huang et al. [2014] added orthogonal constraints to eq (4.13) to fix scale transfer problems. Similar to before, we have the following K subproblems:

$$\min \quad \left\| \mathbf{X}_k - \mathbf{F}_k \mathbf{S}_k \mathbf{G}_k^T \right\|_F^2 - \alpha_k \mathbf{tr}\left(\mathbf{F}_k^T \mathbf{A}_k \mathbf{F}_k \right)$$
$$- \left(\mathbf{G}_k^T \mathbf{A}_k' \mathbf{G}_k \right)$$
$$\text{s.t.} \quad \mathbf{F}_k \geq 0, \ \mathbf{S}_k \geq 0, \ \mathbf{G}_k \geq 0, \ k = 1, \ldots, K$$
$$\mathbf{F}_k^T \mathbf{D}_k \mathbf{F}_k = \mathbf{I}, \mathbf{G}_k^T \mathbf{D}_k' \mathbf{G}_k = \mathbf{I} \tag{4.14}$$

where $\mathbf{A}_k'$ is the keyword-keyword adjacency matrix, $\mathbf{D}_k' = \sum_j \mathbf{A}_k'^{ij}$ is the degree matrix, $\mathbf{I}$ is the identity matrix of the appropriate size. The main difference between

GNMF eq (4.13) and IGNMTF eq (4.14) is the orthogonal constraints, which fix both the scale transfer and trivial solution problems. Without the constraints the optimization problem in eq (4.13) can be seen as a special case of eq (4.14) by absorbing $\mathbf{S}_k$ into $\mathbf{F}_k$. Also, as shown in Nie et al. [2010] when orthonormal and non-negative constraints of $\mathbf{F}_k$ and $\mathbf{G}_k$ are simultaneously satisfied, then it can be proved that in each row of $\mathbf{F}_k$ and $\mathbf{G}_k$, only one element could be positive and others are zeros, which can be directly used to assign cluster labels to data points.

4.2.2.2 Proposed Framework

As shown in the last subsubsection, existing work on NMTF is designed for a single social network, and cannot be readily applied to model multiple social networks and identify similar actors. Notice that disease-specific social networks often share the same set of topics. For example, for diabetes-specific social networks, the set of topics usually include Type I diabetes, Type II diabetes, gestational diabetes, diet and exercise, etc. Furthermore, the users of these social networks tend to form the same groups with interest in certain topics. For example, on both TuDiabetes and DiabetesSisters, there are user groups associated with Type I diabetes, Type II diabetes and gestational diabetes. Based on this observation, in this subsection, we present our proposed optimization framework named CrossNet, which jointly decomposes the user-keyword matrices from multiple social networks, while requiring them to share the same topics as well as user group-topic association matrices.

$$
\begin{aligned}
\min \quad & \sum_{k=1}^{K} \left\{ \left\| \mathbf{X}_k - \mathbf{F}_k \mathbf{S} \mathbf{G}^T \right\|_F^2 + \alpha_k tr\left(\mathbf{F}_k^T \mathbf{L}_k^F \mathbf{F}_k \right) \right\} + \beta_k tr\left(\mathbf{G}^T \mathbf{L}_k^G \mathbf{G} \right) \right\} \\
\text{s.t.} \quad & N_F(\mathbf{F}_k), N_G(\mathbf{G}), N_S(\mathbf{S}) \\
& O_F(\mathbf{F}_k), O_G(\mathbf{G}), k = 1, \dots, K
\end{aligned}
\tag{4.15}
$$

where $\mathbf{L}_k^F = \mathbf{I} - \mathbf{D}_k^{-\frac{1}{2}} \mathbf{A}_k \mathbf{D}_k^{-\frac{1}{2}}$ is the symmetric normalized Laplacian of the user-user adjacency matrix $\mathbf{A}_k$, $\mathbf{L}_k^G = \mathbf{I} - \mathbf{D}_k^{-\frac{1}{2}} \mathbf{A}_k^G \mathbf{D}_k^{-\frac{1}{2}}$ is the symmetric normalized Laplacian of the keyword-keyword adjacency matrix $\mathbf{A}_k^G$, $N_F(\cdot)$, $N_G(\cdot)$, and $N_S(\cdot)$ denote the non-negative constraint on a certain matrix, $O_F(\cdot)$ and $O_G(\cdot)$ denote the orthogonal constraint on the input matrix. Notice that we use the symmetric normalized Laplacian as it provides more robust results as compared to the one used in eq (4.13).

Compared with eq (4.13) and eq (4.14), the major difference is that we couple the K subproblems by requiring them to share the same matrices $\mathbf{S}$ and $\mathbf{G}$. This is because multiple disease-specific social networks tend to share the same topics ($\mathbf{G}$) as well as the user group-topic matrix $\mathbf{S}$. Depending on the specific form of the non-negative constraint $N(\cdot)$ and the orthogonal constraint $O(\cdot)$, CrossNet can be instantiated in four different ways as follows.
CrossNet-I:

$$
\mathbf{F}_k \geq 0, \mathbf{G} \geq 0
$$
$$
\mathbf{F}_k^T \mathbf{F}_k = \mathbf{I}_F, \sum_j \mathbf{G}_{i,j} = 1, k = 1, \dots, K.
\tag{4.16}
$$

CrossNet-II:

$$\mathbf{F}_k \geq 0, \mathbf{S} \geq 0, \mathbf{G} \geq 0$$
$$\mathbf{F}_k^T \mathbf{F}_k = \mathbf{I}_F, \sum_j \mathbf{G}_{i,j} = 1, k = 1, \dots, K. \tag{4.17}$$

CrossNet-III:

$$\mathbf{F}_k \geq 0, \mathbf{G} \geq 0$$
$$\mathbf{F}_k^T \mathbf{D}_F \mathbf{F}_k = \mathbf{I}_F, \sum_j \mathbf{G}_{i,j} = 1, k = 1, \dots, K. \tag{4.18}$$

CrossNet-IV:

$$\mathbf{F}_k \geq 0, \mathbf{S} \geq 0, \mathbf{G} \geq 0$$
$$\mathbf{F}_k^T \mathbf{D}_F \mathbf{F}_k = \mathbf{I}_F, \sum_j \mathbf{G}_{i,j} = 1, k = 1, \dots, K. \tag{4.19}$$

Notice that in all four instantiations, the orthogonal constraint on $\mathbf{G}$ is designed in such a way that its row sums are equal to 1. In this way, we allow the keywords to be part of multiple keyword groups (topics) instead of a single one.

4.2.2.3 Optimization Algorithm

In this subsubsection we provide the optimization algorithm for CrossNet with the constraint instantiation in eq (4.19). The algorithm for the other instantiations can be designed in a similar way. The objective function in eq (4.15) that we minimize is the following sum of squared residuals:

$$f = \sum_{k=1}^{K} \left\{ \text{tr}\left(\mathbf{X}_k^T \mathbf{X}_k - 2\mathbf{G}^T \mathbf{X}_k^T \mathbf{F}_k \mathbf{S} + \mathbf{F}_k^T \mathbf{F}_k \mathbf{S} \mathbf{G}^T \mathbf{G} \mathbf{S}^T \right) \\ + \alpha_k \text{tr}\left(\mathbf{F}_k^T \mathbf{L}_k^F \mathbf{F}_k \right) \right\}$$

Following the standard theory of constrained optimization, we introduce the following Lagrangian function where Lagrange multiplier Λ_k enforce the constraints $\mathbf{F}_k^T \mathbf{D}_k \mathbf{F}_k = \mathbf{I}$ in eq (4.19).

$$\mathcal{L} = \sum_{k=1}^{K} \left\{ \text{tr}\left(\mathbf{X}_k^T \mathbf{X}_k - 2\mathbf{V}^T \mathbf{X}_k^T \mathbf{F}_k \mathbf{S} + \mathbf{F}_k^T \mathbf{F}_k \mathbf{S} \mathbf{G}^T \mathbf{G} \mathbf{S}^T \right) \\ + \alpha_k \text{tr}\left(\mathbf{F}_k^T \mathbf{L}_k^F \mathbf{F}_k \right) + \Lambda_k \left(\mathbf{I} - \mathbf{F}_k^T \mathbf{D}_k \mathbf{F}_k \right) \right\} \tag{4.20}$$

Computing $\mathbf{F}_k$: Fixing $\mathbf{S}$ and $\mathbf{G}$, the gradient $\nabla \mathcal{L}(\mathbf{F}_k)$ is

$$\nabla \mathcal{L}(\mathbf{F}_k) = 2(\mathbf{F}_k \mathbf{S} \mathbf{G}^T \mathbf{G} \mathbf{S}^T + \alpha_k \mathbf{L}_k^F \mathbf{F}_k - \mathbf{X}_k \mathbf{G} \mathbf{S}^T - \mathbf{D}_k \mathbf{F}_k \Lambda_k)$$

By the KKT complementary slackness we have $\nabla \mathcal{L}(\mathbf{F}_k)^{ij} \mathbf{F}_k^{ij} = 0$, so

$$(\mathbf{F}_k \mathbf{S} \mathbf{G}^T \mathbf{G} \mathbf{S}^T + \alpha_k \mathbf{L}_k^F \mathbf{F}_k - \mathbf{X}_k \mathbf{G} \mathbf{S}^T - \mathbf{D}_k \mathbf{F}_k \Lambda_k)^{ij} \mathbf{F}_k^{ij} = 0$$

The Lagrangian multiplier Λ^k is calculated as given in the Ding et al. [2006] by summing up across i index. That gives

$$\Lambda_k = \mathbf{F}_k^T \mathbf{X}_k \mathbf{G} \mathbf{S}^T - \mathbf{S} \mathbf{G}^T \mathbf{G} \mathbf{S}^T - \alpha_k \mathbf{F}_k^T \mathbf{L}_k^F \mathbf{F}_k$$

As Λ_k has negative components, it can be expressed as a difference of two non-negative components $\Lambda_k = \Lambda_k^+ - \Lambda_k^-$, where $\Lambda_k^+ = \frac{|\Lambda_k| + \Lambda_k}{2}$ and $\Lambda_k^- = \frac{|\Lambda_k| - \Lambda_k}{2}$. Substituting the non-negative components in eq 4.2.2.3 we get

$$(\mathbf{F}_k \mathbf{S} \mathbf{G}^T \mathbf{G} \mathbf{S}^T + \alpha_k \mathbf{L}_k^F \mathbf{F}_k - \mathbf{X}_k \mathbf{G} \mathbf{S}^T - \mathbf{D}_k \mathbf{F}_k \Lambda_k^+$$
$$+ \mathbf{D}_k \mathbf{F}_k \Lambda_k^-)^{ij} \mathbf{F}_k^{ij} = 0$$

As the constraint, $\mathbf{I} - \mathbf{F}_k^T \mathbf{D}_k \mathbf{F}_k$ is symmetric, as suggested in Gu et al. [2011] we have $\mathbf{tr}(\Lambda_k(\mathbf{I} - \mathbf{F}_k^T \mathbf{D}_k \mathbf{F}_k)) = tr((\mathbf{I} - \mathbf{F}_k^T \mathbf{D}_k \mathbf{F}_k)\Lambda_k^T)$. Therefore only the symmetric part of Λ_k contributes to $\mathscr{L}$. So Λ_k should be symmetric, we use $\Lambda'_k = \frac{\Lambda_k + \Lambda_k^T}{2}$ instead of Λ_k. This leads to the following update rule for calculating $\mathbf{F}_k$:

$$\mathbf{F}_k^{ij} \Leftarrow \mathbf{F}_k^{ij} \sqrt{\frac{\left\{ \mathbf{X}_k \mathbf{G} \mathbf{S}^T + \mathbf{D}_k \mathbf{F}_k \Lambda_k'^+ \right\}^{ij}}{\left\{ \mathbf{F}_k \mathbf{S} \mathbf{G}^T \mathbf{G} \mathbf{S}^T + \alpha_k \mathbf{L}_k^F \mathbf{F}_k + \mathbf{D}_k \mathbf{F}_k \Lambda_k'^- \right\}^{ij}}} \qquad (4.21)$$

Computing G: Fixing $\mathbf{S}$ and $\mathbf{F}_k$, setting $\nabla \mathscr{L}(\mathbf{G}) = 0$ and following the similar steps in computing $\mathbf{F}_k$ we get the following update rule for $\mathbf{G}$:

$$\mathbf{G}^{ij} \Leftarrow \mathbf{G}^{ij} \sqrt{\frac{\left\{ \sum_{t=1}^T \mathbf{X}_k^T \mathbf{F}_k \mathbf{S} \right\}^{ij}}{\left\{ \sum_{t=1}^T \mathbf{S}^T \mathbf{F}_k^T \mathbf{F}_k \mathbf{S} \mathbf{G} \right\}^{ij}}} \qquad (4.22)$$

The orthogonal constraint $\sum_j \mathbf{G}_{i,j} = 1$ on $\mathbf{G}$ is enforced by row normalizing the $\mathbf{G}$ factor after every iteration.

Computing S: Fixing $\mathbf{G}$ and $\mathbf{F}_k$, setting $\nabla \mathscr{L}(\mathbf{S}) = 0$ and following the similar steps in computing $\mathbf{F}_k$ we get the following update rule for $\mathbf{S}$:

$$\mathbf{S}^{ij} \Leftarrow \mathbf{S}^{ij} \sqrt{\frac{\sum_{k=1}^K \left\{ \mathbf{F}_k^T \mathbf{X}_k \mathbf{G} \right\}^{ij}}{\sum_{k=1}^K \left\{ \mathbf{F}_k^T \mathbf{F}_k \mathbf{S} \mathbf{G}^T \mathbf{G} \right\}^{ij}}} \qquad (4.23)$$

Theorem 4.2

The objective function in eq (4.16) is lower-bounded, and monotonically decreasing (non-increasing) with the update rules eq (4.21), eq (4.22) and eq (4.23). Hence CrossNet converges.   ∎

Proof Sketch. First of all, it is easy to see that the objective function in eq (4.16) is lower-bounded. Second, it consists of two terms, and it suffices to show that each of these terms is monotonically decreasing. As the second term depends on **U** only, the update functions are similar between CrossNet and general NMTF. Following the steps in Ding et al. [2006], Gu et al. [2011], it can be shown that the first term is monotonically decreasing under the update rules. For the second term, by introducing an auxiliary function as in Cai et al. [2011], it can be shown that the second term is also monotonically decreasing. Putting everything together, the update rules converge to the local optimal solution. Hence CrossNet converges. Details are omitted due to the space limit. ∎

With the update rules eq (4.21), eq (4.22) and eq (4.23) the optimization algorithm for link prediction problem is presented in Algorithm 2.

Algorithm 2: CrossNet Algorithm

    Input: A set of user-user adjacency matrices $\{\mathbf{A}_1, \mathbf{A}_2, \ldots, \mathbf{A}_K\}$ constructed
             from user relationship graphs $\mathscr{G}_k, k = 1, \ldots, K$ and a set of
             user-keyword adjacency matrices $\{\mathbf{X}_1, \mathbf{X}_2, \ldots, \mathbf{X}_K\}$ constructed from
             user-keyword bipartite graphs $\mathscr{G}'_k, k = 1, \ldots, K$. The regularization
             parameter α^k. Number of iterations t.
    Output: The user latent factors $\mathbf{F}_k$ for all the disease-specific social
             networks $k = 1, \ldots, K$.
1 Initialize the factor matrices $\mathbf{F}_k$ and $\mathbf{G}$ using k-means.
2 **for** $i \leftarrow 1$ *to* t **do**
3 | Update $\mathbf{S}$ using eq (4.23)
4 | Update $\mathbf{G}$ using eq (4.22)
5 | Update $\mathbf{F}_k$ using eq (4.21) $\forall\, k = 1, \ldots, K$
6 **end**
7 Return user latent factors $\mathbf{F}_k$.

4.2.2.4   Link Recommendation

Using NMTF we represent the users in a latent feature space shared across all the networks. For link prediction we leverage the learned shared user space along with user associations in each social network. We combine user-user associations and user-user latent features space as a graph. We use neighborhood formation using random walk with restarts (RWR) [Sun et al., 2005a] to learn the cross network user-user relations. As the social networks are dynamic in nature (users join and leave over time), our approach is more robust and works for new users as we can leverage user-user associations to predict links between cross network users.

4.2.2.5   Complexity Analysis

The user-keyword matrix $\mathbf{X} \subset \mathbb{R}^{m \times n}$ is typically very sparse . Using NMTF, $\mathbf{X}$ is factorized into three latent factors as shown in eq (4.12). Updating $\mathbf{F}_k$, $\mathbf{S}$ and $\mathbf{G}$ using a multiplicative update algorithm takes $O(k^2(m+n))$ in each iteration for computation, and other $O(zk)$ cost for component wise addition where $z \ll mn$ is the number of non-zero elements in **X**. Using the multiplicative algorithms for sparse computation, the efficiency of our algorithm can be improved tremendously. As the value of k is very small (usually < 100), we can consider that the algorithm is linear per computation. Empirically we found that the number of iterations it takes to converge is $t < 100$. So the total cost of complexity is $O(tk^2(m+n)+tkz)$ which is still linear. So computationally, CrossNet scales to large data sets.

4.2.3 RESULTS

In this section we compare CrossNet with other state-of-the-art approaches on an academic publications data set. We also demonstrate the effectiveness of CrossNet through a case study on a diabetes-specific social network data set.

4.2.3.1 Data Sets

The first data set is from the online repository of electronic preprints - *arXiv*, which contains scientific papers related to artificial intelligence (cs.AI), computer vision (cs.CV), databases (cs.DB), machine learning (cs.LG) and software (cs.SE) categories in the field of computer science. Each category represents a social network with user-user associations based on the co-authorship information. Keywords are extracted from the abstract of each scientific paper. For each author (user), we combine all the abstracts from the papers authored or co-authored by the author. The ground truth for this data set is computed from the existing cross network links (authors common to different networks). The neighborhood formation algorithm based on RWR is used to estimate the cross network link associations.

Table 4.4

Statistics of Arxiv and Diabetes-Specific Social Network Data Sets

ArXiv data set	# papers	# nodes	# edges
Artificial Intelligence (cs.AI)	6972	10272	31266
Computer Vision (cs.CV)	5321	10156	19284
Databases (cs.DB)	2070	4297	6492
Machine Learning (cs.LG)	7321	11103	39349
Software (cs.SE)	2753	5514	18462
Diabetes data set	# posts	# nodes	# edges
Diabetes Sisters	2643	750	4118
TuDiabetes	3742	1032	6323

We also demonstrate the applicability of CrossNet to a real world setting through a case study on diabetes-specific social networks. The user posts are crawled from two diabetes-specific social networks – TuDiabetes and Diabetes Sisters. The user-user associations in the forums are missing, so we consider the users who post in any given thread as related, i.e., there exists an edge between the users responding to the same thread. Keywords are extracted from the posts. Several pre-processing steps were taken before the experiments, including stemming, stop word removal, etc. Each user is represented as a binary feature vector with a bag of words with n-grams $n = \{1,2,3\}$. Table 4.4 shows the data set statistics.

Table 4.5

Arxiv Results

	DB - SE		DB - LG		LG - SE		AI - LG		AI - CV		CV - LG	
	P@10	P@20	P@10	P@20	P@10	P@20	P@10	P@20	P@10	P@20	P@10	P@20
GNMF	15.48	12.84	12.88	12.77	8.64	8.77	18.72	17.47	15.03	14.9	11.73	11.9
IGNMTF	26.28	18.36	18.92	16.3	24.25	18.96	33.23	30.26	22.08	19.02	32.9	25.73
CoupledLP	23.4	20.28	31.58	21.77	24.64	24.12	37.86	40.87	36.84	25.4	33.43	32.73
COSNET	31.68	29.88	35.8	26.56	31.99	28.64	**47.58**	**45.86**	41.76	30.99	43.4	38.85
CrossNet - I	35.28	30.48	36.71	32.95	33.8	31.48	44.62	42.73	42.83	38.44	45.85	42.7
CrossNet - II	**36.12**	30.59	36.94	33.29	34.06	31.61	44.77	42.32	43.09	38.84	46.2	42.88
CrossNet - III	35.28	30.36	36.59	33.17	33.93	31.48	44.77	42.54	42.69	38.7	46.03	42.7
CrossNet - IV	35.41	**30.63**	**37.05**	**33.63**	**34.19**	**31.73**	45.08	42.81	**43.23**	**39.24**	**46.38**	**43.05**

Table 4.6

Diabetes Keyword Groups (Top 7). $< .. >^1$ Represents Keywords from Diabetes Sisters, $< .. >^2$ from Tudiabetes and $< .. >^{12}$ from Both

topic-1 healthy eating	topic-2 insurance	topic-3 exercise	topic-4 products	topic-5 diet	topic-6 diagnosis	topic-7 research
food[12]	medical insurance[12]	running[12]	pump[12]	insulin[12]	diagnosed[12]	patients study[1]
healthy eating[12]	cost information[12]	ginger[2]	cgm[12]	dose[12]	diabetes[12]	levels[12]
carbs[12]	money[12]	training[1]	minimed[12]	carbs[12]	family doctor[12]	doctor[12]
protein[2]	insulin supplies[12]	yoga[12]	infusion pumps[1]	low carb[12]	hospital[12]	ADA[1]
veggies[1]	strips[2]	gym[12]	insulin use[12]	high day[2]	symptoms[12]	people[12]
bread[12]	companies[12]	workout[12]	omnipod[12]	bg[12]	months[12]	clinical treatment[1]
diet[12]	doctors[12]	muscle[2]	pumping set[12]	basal hours[12]	told diabetic[12]	disease research[2]

4.2.3.2 Experiment Setup

We compare the proposed CrossNet approaches with other state-of-the-art approaches including: (1) GNMF [Cai et al., 2011]; (2) IGNMTF [Gu et al., 2011]; (3) CoupledLP [Dong et al., 2015] modified for cross network links; and (4) COS-NET [Zhang et al., 2015].

We have used Precision at K (P@K) as an evaluation metric to compare the performance of different algorithms. It computes the percentage of the relevant links among the top-K links predicted by the algorithm. For evaluation we compute P@10 and P@20 for all the algorithms and data set combinations. Here relevant links refer to the links between similar actors across the networks.

Regarding the parameters, we use grid-search to set regularization parameters $\alpha_1 = \alpha_2 = 0.01$ for CrossNet, the number of user groups and keyword groups $o = p = 40$ and iterations $t = 100$. From the results in Table 4.5 CrossNet outperforms all other approaches. Jointly factorizing keywords across all the networks through **G** resulted in significant improvement over GNMF and IGNMF approaches. CrossNet outperformed modified CoupledLP as it uses both the user-user associations and user-keyword bipartite graphs unlike CoupledLP that relies on user-user network structure only. COSNET performs closely as it leverages both the user-user and user-keyword graphs, but it identifies the distinct user-user links across networks to the similar ones. Among the four constraint instantiations, setting **S** ≥ 0 and an orthogonal constraint with degree matrix led to a better performance.

4.2.3.3   Case Study

We also conduct a case study on diabetes-specific social networks. Notice that Cross-Net has two steps: (1) jointly decomposing the user-keyword matrices from each network into respective user factors and a combined keyword latent factor matrix; (2) using RWR on user-user associations and user factor matrices for each network to recommend links between similar actors across different networks. Table 4.6 shows the keyword latent factors from all the networks combined ($K = 2, p = 7$). It can be observed that our joint factorization approach clustered similar keywords from different networks into one group. The following is an example of two posts generated by two users from different social networks, between whom CrossNet recommends a link.

User A: I have been diagnosed with Type 1 for about 5 years. I had my blood glucose with an A1C over 9. I am worried! *User B: I am a 22 year old female recently diagnosed type 1 diabetic. I found out that my blood glucose was over 400. I came here looking for support.*

As we can see, both users are concerned about their blood glucose level and have been diagnosed with Type I diabetes.

4.3   SOURCE-FREE DOMAIN ADAPTATION

In the past decades, the advancements in the field of machine learning have led to their wide adoption to solve different real world applications. In general, training a new machine learning model needs a large amount of labeled data. In some applications, such large-scale annotated data sets are readily available, giving rise to an increasing number of off-the-shelf tools. For example, the *Caffe Model Zoo* hosts different models that can be readily used for various classification tasks; language processing tools such as *Stanford NLP Toolkit* [Manning et al., 2014] come with various models for natural language processing tasks. However, many of these machine learning tasks are time-evolving in nature due to the emergence of new features and the shift in class conditional distribution. As a result, the off-the-shelf tools may not be able to adapt to such changes in a timely fashion, and will suffer from sub-optimal performance in the learning task.

On the other hand, existing work on transfer learning, cannot be readily applied to improve the performance of the off-the-shelf tools due to the lack of the training data for obtaining these tools, i.e., the lack of source domain data. More specifically, due to licensing or other copyright restrictions, the labeled data sets sometimes are not released but the underlying models are made available to use as a black-box classifier [Viola and Jones, 2001]. Therefore, given these black box classifiers, *is it possible to leverage these classifiers to improve the classification performance on the evolved source domain, i.e., the target domain, given a limited amount of training data from the target domain?* To this end, we are facing two major challenges: (1) label deficiency happens when new features appear, or the relationship between individual features and the class labels changes in the target domain; (2) distribution shift happens when the class conditional distribution in the target domain is different from the training data used by off-the-shelf classifiers, potentially changing the optimal predicted label.

In this paper we address the above mentioned challenges of label deficiency and distribution shift through the proposed Adaptive Off-The-shelf classification (**AOT**) framework. In our case, we consider that there exists a black-box classifier that gives out the classification labels for the target domain examples, and no other information about the black-box is known. In particular, we assume that the training data used to obtain the off-the-shelf classifiers are not available, and we aim to leverage the noisy class labels predicted by the black-box classifier and very few labeled examples from the target domain to improve the classification performance of the unlabeled data in the target domain. Given an unlabeled document from the target domain, these tools are able to predict the polarity of the text without taking into consideration the unique characteristics of this domain. The proposed framework is able to effectively integrate the information from these tools as well as the few labeled examples from the target domain to construct a classification model for the target domain with significantly improved performance.

The following are the main contributions of this section; (1) a novel problem setting of source free domain adaptation, where the goal is to leverage the output of an off-the-shelf classifier and a few labeled examples from the target domain, in order to obtain a significantly better classification model for the target domain, as compared to the off-the-shelf classifier; (2) a generic optimization framework named **AOT** to adapt an off-the-shelf classifier to the target domain by explicitly addressing the two major types of changes from the source domain to the target domain, i.e., label deficiency and distribution shift; (3) analysis of the performance of the proposed **AOT** framework in terms of convergence to the global optimum, and the complexity of the proposed algorithm.

4.3.1 PROBLEM DEFINITION

In this section, we introduce the notation used in the paper and formally define the problem of source free domain adaptation with an off-the-shelf classifier. Let

$\mathscr{D}_L = \{(\mathbf{x}_i, y_i)\}_{i=1}^m$ be the set of m labeled examples from the target domain; $\mathscr{D}_U = \{(\mathbf{x}_i)\}_{i=m+1}^{m+n}$ be the set of n unlabeled examples from the target domain, where $\mathbf{x}_i \in \mathbb{R}^d$ is a real valued vector of size d; and $y_i \in \{-1, 1\}, \forall i \in 1, \ldots, m$ be the binary class label. We consider the number of labeled examples to be much smaller than the number of unlabeled examples, i.e., $m \ll n$. Let $\mathbf{f}^0 = [y_1^0, \ldots, y_{m+n}^0]^T$ be a $(m+n)$-dimensional vector consisting of the pseudo-labels generated by the off-the-shelf classifier, where $y_i^0 \in \{-1, 1\}, i \in 1, \ldots, m+n$, and $c_i \in [0, 1), i = 1, \ldots, m+n$, be the confidence score for each of the $m+n$ examples ($\mathscr{D}_L \cup \mathscr{D}_U$).

First of all, we represent all $m+n$ examples from the target domain as a graph $G = (V, E)$, where V is the set of nodes, and E is the set of edges. In this graph, each node corresponds to an example, labeled or unlabeled, i.e., $|V| = m+n$, and the weight associated with each edge measures the similarity between a pair of nodes. Let $\mathbf{W}$ be the affinity matrix of this graph, whose non-negative element $\mathbf{W}_{ij}$ in the ith row and j^{th} column is the weight of the edge connecting the examples $\mathbf{x}_i$ and $\mathbf{x}_j$. Let $\mathbf{D}$ be the $(m+n) \times (m+n)$ degree matrix whose diagonal elements are set to be $\sum_j \mathbf{W}_{ij}$. The normalized Laplacian of the affinity matrix $\mathbf{W}$ is given by $\mathbf{S} = \mathbf{D}^{-\frac{1}{2}} \mathbf{W} \mathbf{D}^{-\frac{1}{2}}$.

The problem of source free domain adaptation is to adapt the noisy pseudo-labels $\mathbf{f}^0$ from the off-the-shelf classifier to the examples in the target domain by leveraging the information of a small number of labeled examples $\mathscr{D}_L$ from the target domain, without having access to the source domain data based on which the off-the-shelf classifier was trained. More specifically, given a set of m labeled examples $\mathscr{D}_L$; the n unlabeled examples $\mathscr{D}_U$; the normalized affinity matrix for the $m+n$ examples $\mathbf{S}$; and the noisy pseudo-labels $\mathbf{f}^0$ from the off-the-shelf classifier; the goal of source free domain adaptation is to learn a classification vector $\mathbf{f} \in \mathbb{R}^{m+n}$ to correctly classify all the $m+n$ examples. Notice that unlike $\mathbf{f}^0$, the elements of $\mathbf{f}$ may not be binary. Therefore, the predicted class label $\hat{\mathbf{y}}_i$ of the unlabeled examples $\mathscr{D}_U$ is set as $\hat{\mathbf{y}}_i = +1$, if $f_i \geq 0, i \in m+1, \ldots, m+n$, and $\hat{\mathbf{y}}_i = -1$ otherwise, where f_i is the ith element of $\mathbf{f}$.

4.3.2 PROPOSED APPROACH

In this section, we propose our **AOT** framework. The goal of the **AOT** framework is to learn the classification vector $\mathbf{f}$ for all the $m+n$ examples based on $\mathbf{f}^0$. Usually the pseudo-labels from the black-box classifier $\mathbf{f}^0$ are noisy due to label deficiency and distribution shift. As shown in eq. (4.24), we decompose the classification vector $\mathbf{f}$ into the sum of noisy pseudo-labels from off-the-shelf classifier $\mathbf{f}^0$ and two residual vectors:

$$\mathbf{f} = \mathbf{f}^0 + \Delta_1 \mathbf{f} + \Delta_2 \mathbf{f} \tag{4.24}$$

where $\Delta_1 \mathbf{f} \in \mathbb{R}^{m+n}$ and $\Delta_2 \mathbf{f} \in \mathbb{R}^{m+n}$ are the residual vectors that address label deficiency and distribution shift respectively. More specifically, the residual vector $\Delta_1 \mathbf{f}$ accounts for the change in the relationship between features and class labels in the target domain. For example, in sentiment classification, with the emergence of new words in the target domain, $\Delta_1 \mathbf{f}$ will provide correcting information regarding the relationship between the new words and the class labels. On the other hand, the residual vector $\Delta_2 \mathbf{f}$ addresses the changes in class conditional distribution in the target domain compared to the source domain data used to train the off-the-shelf classifier. For example, in sentiment classification, $\Delta_2 \mathbf{f}$ will provide insights regarding the potentially different sentiment polarity for a certain combination of keywords that are specific to the target domain.

In our framework, we propose to solve for both residual vectors via the following generic optimization problem.

$$Q(\Delta_1 \mathbf{f}, \Delta_2 \mathbf{f}) = Q_1(\Delta_1 \mathbf{f}, \Delta_2 \mathbf{f}) + Q_2(\Delta_1 \mathbf{f}) + Q_3(\Delta_2 \mathbf{f}) \tag{4.25}$$

where Q_1 takes into consideration both residual vectors, and it aims to enforce label consistency on all the examples along the data manifold; Q_2 is a sparsity constraint on the label deficiency residual vector $\Delta_1 \mathbf{f}$; and Q_3 is the objective function of $\Delta_2 \mathbf{f}$ for addressing the distribution shift. The optimal residual vectors $(\Delta_1 \mathbf{f}^*, \Delta_2 \mathbf{f}^*)$ are computed as follows:

$$(\Delta_1 \mathbf{f}^*, \Delta_2 \mathbf{f}^*) = \underset{\Delta_1 \mathbf{f} \in \mathbb{R}^{m+n}, \Delta_2 \mathbf{f} \in \mathbb{R}^{m+n}}{\operatorname{argmin}} Q(\Delta_1 \mathbf{f}, \Delta_2 \mathbf{f}) \tag{4.26}$$

To solve this optimization problem, we propose to use the alternating minimization strategy. More specifically:

$$\Delta_1 \mathbf{f}_{t+1} = \underset{\Delta_1 \mathbf{f} \in \mathbb{R}^{m+n}}{\operatorname{argmin}} Q_1(\Delta_1 \mathbf{f}, \Delta_2 \mathbf{f}_t) + Q_2(\Delta_1 \mathbf{f}) \tag{4.27}$$

$$\Delta_2 \mathbf{f}_{t+1} = \underset{\Delta_2 \mathbf{f} \in \mathbb{R}^{m+n}}{\operatorname{argmin}} Q_1(\Delta_1 \mathbf{f}_t, \Delta_2 \mathbf{f}) + Q_3(\Delta_2 \mathbf{f}) \tag{4.28}$$

where $t = 0, \ldots, T-1$, T is the total number of iterations, and $\Delta_1 \mathbf{f}_t$ ($\Delta_2 \mathbf{f}_t$) is the vector $\Delta_1 \mathbf{f}$ ($\Delta_2 \mathbf{f}$) in the tth iteration. The proposed **AOT** algorithm (Algo. 3) runs until convergence or the max number of iterations is reached. It takes the normalized affinity matrix $\mathbf{S}$ and the set of noisy class labels generated by the off-the-shelf classifier $\mathbf{f}^0$ as input, and outputs the classification vector $\mathbf{f}$ for the examples in the target domain. After the initialization step (Step 1), the algorithm iteratively updates $\Delta_1 \mathbf{f}$ and $\Delta_2 \mathbf{f}$ in Steps 3 and 4. After T iterations or convergence, the **AOT** algorithm outputs the vector $\mathbf{f}$ for all the examples in the target domain. As discussed earlier, the predicted class label $\hat{\mathbf{y}}_i$ of the unlabeled examples $\mathscr{D}_U$ is set as $\hat{\mathbf{y}}_i = +1$, if $f_i \geq 0, i \in m+1, \ldots, m+n$, and $\hat{\mathbf{y}}_i = -1$ otherwise, where f_i is the ith element of $\mathbf{f}$.

Next, we introduce the proposed techniques for computing the residual vectors $\Delta_1 \mathbf{f}$ and $\Delta_2 \mathbf{f}$ in Subsections 4.3.2.1 and 4.3.2.2 respectively, the convergence analysis of the proposed **AOT** framework in Subsection 4.3.2.3.

4.3.2.1 Label Deficiency

In this subsubsection, we introduce our proposed techniques to solve for residual vector $\Delta_1 \mathbf{f}$, which addresses label deficiency. Based on eq. (4), this involves the minimization of both Q_1 and Q_2.

Algorithm 3: AOT : Adaptive Off-the-Shelf Classifier

Input: (1) The normalized affinity matrix $\mathbf{S}$ for the $m+n$ examples in the target domain; (2) the noisy class labels $\mathbf{f}^0$ generated by the off-the-shelf classifier; (3) the max number of iterations T

Output: f: The classification vector for all the examples in the target domain.

1 Initialize $\Delta \mathbf{f}_0 = \mathbf{0}^{m+n}$
2 **for** $t = 1$ *to* T **do**
3 | Fix $\Delta_1 \mathbf{f}$, compute $\Delta_2 \mathbf{f}$ using **ADDRESSLABDEF**
4 | Fix $\Delta_2 \mathbf{f}$, compute $\Delta_1 \mathbf{f}$ using **ADDRESSDISSHIFT**
5 **end**
6 **return** $\mathbf{f} = \mathbf{f}^0 + \Delta_1 \mathbf{f} + \Delta_2 \mathbf{f}$

To instantiate Q_1, notice that the key to semi-supervised learning is the *consistency* assumption [Zhou et al., 2003]. When we have access to a small amount of labeled data and lots of unlabeled data, the classification function can be enforced to be sufficiently smooth on the intrinsic structure of the data manifold. According to the consistency assumption, if two examples are similar to each other, they should belong to the same class. So in a scenario where the examples are similar and the corresponding pseudo labels are different, the overall classification vector $\mathbf{f}$ should address the discrepancy in the class labels. More specifically, we have,

$$
\begin{aligned}
Q_1 &= \frac{1}{2} \sum_{i,j=1}^{m+n} \mathbf{W}_{ij} \left(\frac{\mathbf{f}_i^0 + \Delta_1 \mathbf{f}_i + \Delta_2 \mathbf{f}_i}{\sqrt{\mathbf{D}_i}} - \frac{\mathbf{f}_j^0 + \Delta_1 \mathbf{f}_j + \Delta_2 \mathbf{f}_j}{\sqrt{\mathbf{D}_j}} \right)^2 \\
&\quad + \sum_{i=1}^m \mu_1 (\mathbf{f}_i^0 + \Delta_1 \mathbf{f}_i + \Delta_2 \mathbf{f}_i - \mathbf{y}_i)^2 \\
&= \frac{1}{2} (\mathbf{f}^0 + \Delta_1 \mathbf{f} + \Delta_2 \mathbf{f})^T (\mathbf{I} - \mathbf{S})(\mathbf{f}^0 + \Delta_1 \mathbf{f} + \Delta_2 \mathbf{f}) \\
&\quad + \mu_1 \|\mathbf{f}_L^0 + \Delta_1 \mathbf{f}_L + \Delta_2 \mathbf{f}_L - \mathbf{y}_L\|^2
\end{aligned}
\tag{4.29}
$$

where $\mu_1 > 0$ is the regularization parameter. The objective function in eq. (4.29) has two terms. The first term is the smoothness constraint which ensures the class labels of the similar examples are similar to each other. The second term is the regularizer constraint which ensures that the optimal classification function should not change too much from the class labels of the labeled examples.

On the other hand, to instantiate Q_2, we enforce the residual vector $\Delta_1 \mathbf{f}$ to be sparse. The sparsity constraint ensures that this residual vector is non-zero only when the corresponding example contains a changed relationship between features and class labels, or the example has new features. To be specific, we add the elastic-net regularizer to enforce sparsity in the residual vector $\Delta_1 \mathbf{f}$ as follows:

$$
Q_2(\Delta_1 \mathbf{f}) = \mu_2 \|\Delta_1 \mathbf{f}\|_1 + (1 - \mu_2) \|\Delta_1 \mathbf{f}\|_2^2
\tag{4.30}
$$

where μ_2 is the elastic-net coefficient. As the L_1 norm term in the sparse regularizer $Q_2(\Delta_1 \mathbf{f})$ is not continuously differentiable and discontinuous at $\Delta_1 \mathbf{f}_i = 0$, we employ

---

Algorithm 4: ADDRESSLABDEF - Addressing Label Deficiency

Input: (1) The normalized affinity matrix $\mathbf{S}$ for the $m+n$ examples; (2) the noisy pseudo-labels $\mathbf{f}^0$ from the off-the-shelf classifier; (3) the residual vector for distribution shift, $\Delta_2\mathbf{f}$; (4) the max iteration number K

Output: $\Delta_1\mathbf{f}$: The residual vector to address label deficiency.

1   $l_0 = 1, \eta = 2$
2   $\Delta_1\mathbf{f}_0 \leftarrow \mathbf{0}, \gamma_1 = \Delta_1\mathbf{f}_0$ and $t_1 = 1$
3   **for** $k \leftarrow 1$ *to* K **do**
4     $\hat{l} = \eta^i l_{k-1}$
5   **while** $Q_{\text{LABDEF}}(\Delta_1\mathbf{f}, \Delta_2\hat{\mathbf{f}}) > G(prox_{\lambda l}(\gamma_{k-1}), \gamma_{k-1})$ **do**
6       $i \leftarrow i+1$
7       $\hat{l} = \eta^i l_{k-1}$
8   **end**
9     $l_k = \hat{l}; \Delta_1\mathbf{f}_k = prox_{\lambda l}(\gamma_k); t_{k+1} = \frac{1+\sqrt{1+4t_k*t_k}}{2}$
10    $\gamma_{k+1} = \Delta_1\mathbf{f}_k + \frac{t_k-1}{t_{k+1}}(\Delta_1\mathbf{f}_k - \Delta_1\mathbf{f}_{k-1})$
11 **end**
12 **return** $\Delta_1\mathbf{f}_K$

---

the proximal gradient descent [Boyd et al., 2011] to estimate the residual vector $\Delta_1\mathbf{f}$. As shown in eq. (4.27), the combined cost function to address the label deficiency is the sum of the regularization term and the sparsity constraint given as follows:

$$Q_{\text{LABDEF}}(\Delta_1\mathbf{f}, \Delta_2\hat{\mathbf{f}}) = Q_1(\Delta_1\mathbf{f}, \Delta_2\hat{\mathbf{f}}) + Q_2(\Delta_1\mathbf{f}) \tag{4.31}$$

where $\Delta_2\hat{\mathbf{f}}$ is the fixed distribution shift residual vector.

It can be seen that the component Q_1 is a differentiable convex function; detailed proof is omitted due to space constraints. Also, the elastic-net sparsity constraint term Q_2 is closed, convex and non-differentiable over $\Delta_1\mathbf{f}$. The proximal gradient method can be applied to minimize the cost function in eq. (4.31). The proximal gradient step to compute $\Delta_1\mathbf{f}$ is $\Delta_1\mathbf{f}_k = \mathbf{prox}_{t_kQ_2}\left(\Delta_1\mathbf{f}_{k-1} - t_k\nabla Q_2(\Delta_1\mathbf{f})\right)$ where t_k is the step size. For the ith example in the target domain, with elastic-net coefficient μ_2, the proximal mapping for the elastic-net regularizer Q_2 is $\mathbf{prox}_{t_kQ_2}(\mathbf{f}_i) = \left(\frac{1}{\mu_2+2t-2t\mu_2}\right)(\mathbf{f}_i-t)_+ - (-\mathbf{f}_i-t)_-$. The residual vector $\Delta_1\mathbf{f}$ is iteratively computed through proximal gradient descent using a variant of the fast iterative shrinkage thresholding algorithm [Beck and Teboulle, 2009].

The algorithm to address label deficiency is illustrated in Algo. 4. The algorithm takes as input the normalized affinity matrix $\mathbf{S}$ for the $m+n$ examples from the target domain, the noisy pseudo-labels $\mathbf{f}^0$ from the off-the-shelf classifier, the residual vector for distribution shift, $\Delta_2\mathbf{f}$ and the max iteration number K. It outputs the residual vector $\Delta_1\mathbf{f}$ for addressing label deficiency. In the algorithm, we first initialize the parameters, and set the initial label deficiency residual vector to $\Delta_1\mathbf{f}_0 = \mathbf{0}$, a zero vector. The Lipschitz constant for the iteration are computed through line search using the proximal gradient mapping. For any $l > 0$, consider that the proximal gradient mapping at any given point γ is given by $G_l(\Delta_1\mathbf{f}, \gamma) := Q_1(\gamma) + \nabla Q_1(\gamma)^T(\Delta_1\mathbf{f} - \gamma) + \frac{l}{2}||\Delta_1\mathbf{f} - \gamma||^2 + Q_2$ where l is the Lipschitz constant. Considering $\mathbf{L} = \mathbf{I} - \mathbf{S}$, the term $G_l(\Delta_1\mathbf{f}, \gamma)$ can be computed from $Q_1(\Delta_1\mathbf{f})$ and $\nabla Q_1(\Delta_1\mathbf{f})$ terms. In each iteration the Lipschitz constant for the iteration is computed and the residual vector $\Delta_1\mathbf{f}$ is updated through proximal gradient descent steps. As shown in [Beck and Teboulle, 2009], the proposed variant of the fast iterative shrinkage thresholding algorithm (Algo. 4) ensures the cost function Q_{LABDEF} is monotonically decreasing and converges to the global optimal $\Delta_1\mathbf{f}^*$.

4.3.2.2 Distribution Shift

In traditional machine learning, often the data distribution of training and test data is considered to be the same. When the distributions are different, the trained classification model may not perform well on the test data. In source free domain adaptation, the off-the-shelf classifier is trained on a data set with a different distribution from the given data set $\mathcal{D}_L \cup \mathcal{D}_U$ in the target domain. This leads to a distribution shift as the class conditional distribution in the target domain is different from the training data used by off-the-shelf classifiers, potentially changing the optimal predicted labels. The inconsistency in the class labels can be modeled as a residual vector $\Delta_2\mathbf{f}$. Similar to the last subsubsection, the cost function to address the distribution shift is the sum of the regularization term Q_1 and Q_3, which measures the prediction loss on the labeled examples from the target domain:

$$Q_{\text{DISSHIFT}}(\Delta_1\hat{\mathbf{f}}, \Delta_2\mathbf{f}) = Q_1(\Delta_1\hat{\mathbf{f}}, \Delta_2\mathbf{f}) + Q_3(\Delta_2\mathbf{f})$$

$$= Q_1(\Delta_1\hat{\mathbf{f}}, \Delta_2\mathbf{f}) + \frac{1}{m}\sum_i^m \left(y_i - f_i^0 - \Delta_1\hat{f}_i - \Delta_2 f_i\right)^2 \quad (4.32)$$

where $\Delta_1\hat{\mathbf{f}}$ is the fixed label deficiency residual vector. Notice that the cost function Q_{DISSHIFT} is smooth and $\nabla Q_{\text{DISSHIFT}}$ exists for $\Delta_2\mathbf{f} \in \mathbb{R}^{m+n}$. And the term $\nabla Q_{\text{DISSHIFT}}$ can be computed as follows:

$$\nabla Q_{\text{DISSHIFT}} = 2(\mathbf{I} - \mathbf{S})\Delta_2\mathbf{f} + 2\mu_1 \mathbf{I}(\mathbf{f}^0 + \Delta_1\mathbf{f} + \Delta_2\mathbf{f} - \mathbf{y}) \quad (4.33)$$

We employ the gradient boosting approach to compute the residual vector $\Delta_2\mathbf{f}$ that minimizes this cost function. Like other boosting methods, gradient boosting combines a set of *weak learners* into a single strong learner in an iterative fashion. The algorithm for the gradient boosting is shown in Algo. 5. Using the labeled examples, we train a set of gradient boosted regressors and update the residual function $\Delta_2\mathbf{f}$ for all the examples $\Delta_2 f_i = \mathbf{F}(x_i), i \in 1 \ldots (m+n)$ and where $\Delta_2 f_i$ is the ith element of $\Delta_2\mathbf{f}$. The gradient boosted regressor is an ensemble of SVM tree regressors trained on the m labeled examples.

Algorithm **ADDRESSDISSHIFT** (Algo. 5) shows the details for computing the residual vector $\Delta_2\mathbf{f}$ to address the distribution shift. The input to the algorithm are the example feature matrices for the labeled examples $\mathbf{X}_L$, and the unlabeled examples $\mathbf{X}_U$, the noisy pseudo-labels $\mathbf{f}^0$ from the off-the-shelf classifier, the residual vector for label deficiency, $\Delta_1\mathbf{f}$. The gradient boosting is performed by fitting all the labeled

---

Algorithm 5: ADDRESSDISSHIFT - Addressing Distribution Shift

Input: (1) Example feature matrices $\mathbf{X}_L$ and $\mathbf{X}_U$; (2) noisy pseudo-labels $\mathbf{f}^0$
from the off-the-shelf classifier; (3) residual vector for label
deficiency, $\Delta_1 \hat{\mathbf{f}}$; (4) max iterations K
Output: $\Delta_2 \mathbf{f}$: The residual vector to address distribution shift.

1 Initialize $\mathbf{F}_0$
2 **for** $k \leftarrow 1$ *to* K **do**
3      $\mathbf{r}_k = -\nabla Q_{\text{DISSHIFT}}(\Delta \hat{\mathbf{f}}, \Delta_2 \mathbf{f}_k)$
4   Learn a base learner $\mathbf{h}_k$ on labeled examples
5      $\gamma_k = \underset{\gamma}{argmin} \sum_i^m \nabla Q_{\text{DISSHIFT}}(\Delta_1 \hat{f}_i, \mathbf{F}_{k-1} + \gamma \mathbf{h}_k(x_i))$
6      $\mathbf{F}_k = \mathbf{F}_{k-1} + \gamma_k \mathbf{h}_k$
7 **end**
8 **for** $i \leftarrow 1$ *to* $m+n$ **do**
9      $\Delta_2 f_i = \mathbf{F}_k(x_i)$
10 **end**
11 **return** $\Delta_2 \mathbf{f}$

---

examples $\mathscr{D}_L$ to an SVM regressor and the residual value is computed for all the unlabeled examples $\mathscr{D}_U$. Finally, the algorithm outputs the residual vector $\Delta_2 \mathbf{f}$.

4.3.2.3   Convergence of AOT

In this subsection we formally discuss the convergence of the proposed **AOT** algorithm. As discussed earlier in **AOT** algorithm (Algo. 3), we employ an alternative minimization strategy to compute the residual vectors $\Delta_1 \mathbf{f}$ and $\Delta_2 \mathbf{f}$. We follow the existing work [Beck, 2015] to prove the convergence of the proposed alternative minimization framework in Theorem. 4.3.

Theorem 4.3

Let $\Delta_1 \mathbf{f}_k, \Delta_2 \mathbf{f}_k$ be the sequence generated by the proposed alternating minimization based **AOT** framework. Then for any $k > 0, L_1 > 0, L_2 > 0$ and for finite values of L_1 and L_2, the rate of convergence is given by

$$Q(\Delta_1 \mathbf{f}_{k+1}, \Delta_2 \mathbf{f}_k) - Q(\Delta_1 \mathbf{f}^*, \Delta_2 \mathbf{f}^*) \leq ||G_{L_1}^1|| \cdot ||\Delta_1 \mathbf{f}_k + 1 - \Delta_1 \mathbf{f}^*|| \quad (4.34)$$

$$Q(\Delta_1 \mathbf{f}_k, \Delta_2 \mathbf{f}_{k+1}) - Q(\Delta_1 \mathbf{f}^*, \Delta_2 \mathbf{f}^*) \leq ||G_{L_2}^2|| \cdot ||\Delta_2 \mathbf{f}_k + 1 - \Delta_2 \mathbf{f}^*|| \quad (4.35)$$

where $G_{L_1}^1$ and $G_{L_1}^1$ is the proximal gradient mapping, $\Delta_1 \mathbf{f}^*$ and $\Delta_2 \mathbf{f}^*$ are the local optimal residual functions. With the above rate of convergence, the residual functions $\Delta_1 \mathbf{f}_k$ and $\Delta_2 \mathbf{f}_k$ computed iteratively converge to $\Delta_1 \mathbf{f}^*$ and $\Delta_2 \mathbf{f}^*$ respectively.

■

Proof. The cost function for the manifold regularization term $Q_1(\Delta_1\mathbf{f}, \Delta_2\mathbf{f})$ is a continuously differentiable convex function over domain of Q_2, $\mathbb{R}^{m+n}$ and over the domain of Q_3, $\mathbb{R}^{m+n}$. The gradient of Q_1 is (uniformly) Lipschitz continuous with respect to $\Delta_1\mathbf{f}$ over the domain of Q_2 with constant $L_1 \in (0, \infty)$. Also, the gradient of Q_1 is (uniformly) Lipschitz continuous with respect to $\Delta_2\mathbf{f}$ over the domain of Q_3 with constant $L_2 \in (0, \infty)$. Therefore, $||\nabla_1 Q_1(\Delta_1\mathbf{f} + \mathbf{d_1}) - \nabla_1 Q_1(\Delta_1\mathbf{f})|| \leq L_1 ||\mathbf{d_1}||$ and $||\nabla_2 Q_1(\Delta_2\mathbf{f} + \mathbf{d_2}) - \nabla_2 Q_1(\Delta_2\mathbf{f})|| \leq L_2 ||\mathbf{d_2}||$ where the Lipschitz constants $L_1 = L_2 = 2tr((1 + \mu_1)\mathbf{I} - \mathbf{S}) \geq 0$, $\mathbf{d_1} \in \mathbb{R}^{m+n}$, $\mathbf{d_2} \in \mathbb{R}^{m+n}$, $\Delta_1\mathbf{f} + \mathbf{d_1}$ and $\Delta_2\mathbf{f} + \mathbf{d_2}$ is in domain of Q_2 and Q_3 respectively. The proposed alternating minimization framework **AOT** adheres to the framework proposed in the paper [Beck, 2015]. From Lemma 3.4 in the alternating minimization framework proposed in Beck [2015], the sequence $\Delta_1\mathbf{f}_k, \Delta_2\mathbf{f}_k$ generated by the proposed **AOT** framework converges to $\Delta_1\mathbf{f}^*, \Delta_2\mathbf{f}^*$. □

4.3.3   RESULTS

In this subsection we present the experimental results to demonstrate the performance of the proposed **AOT** framework from multiple aspects. We first introduce seven real-world data sets, including four text data sets and three image data sets. Then we compare our **AOT** framework with other state-of-the-art approaches.

Data sets: The performance of the proposed **AOT** framework is evaluated on seven real world data sets. The statistics of all the data sets are shown in Table 4.7. The Stanford sentiment classification tool is used to compute the off-the-shelf classification ratings for all the text data sets. The details of the data sets are as follows:

- **IMDB movie reviews** [Maas et al., 2011a]: A binary sentiment classification data set. The Stanford sentiment classification tool is used to compute the off-the-shelf classification ratings for this data set.
- **Amazon fine food reviews** [McAuley and Leskovec, 2013b]: A binary sentiment classification data set, where all the reviews with 4-5 star ratings are considered as positive and reviews with 1-2 star ratings are considered negative. The Stanford sentiment classification tool is used to compute the off-the-shelf classification ratings for this data set.
- **Convex-nonconvex images** [Erhan, 2007]: A binary image classification data set. The off-the-shelf classifier is on a held out data set with a different SIFT feature set. Figure 4.5 shows images from the data set.
- **Cats and dogs images** [Elson et al., 2007]: A binary image classification data set. The data from Imagenet [Deng et al., 2009] with synsets *cats* and *dogs* are used to train the off-the-shelf classifier. Figure 4.6 shows images from the data set.
- **News articles**: News articles related to illegal immigration and cartel wars in Mexico have been crawled from various news websites from the United States and Mexico. The binary classification task for this data set is to identify whether the content of the news article is related to *illegal immigration* or *cartel wars*. The news articles in Spanish are translated to English

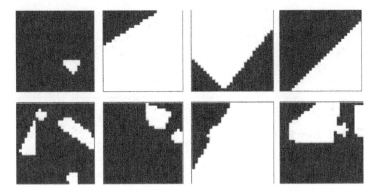

Figure 4.5 Example images from the convex and nonconvex data set.

Figure 4.6 Example images from the cats and dogs data set.

using Google translation service, and used to train the black-box off-the-shelf classifier.

- **Amazon product reviews** [McAuley and Leskovec, 2013a]: A multiclass sentiment classification data set. The reviews ratings to class mappings are as follows: 1-most negative, 2-negative, 3-neutral, 4-positive and 5-most positive. The Stanford sentiment classification tool is used to compute the off-the-shelf classification ratings for this dataset.

- **Office-Caltech dataset** [Hoffman et al., 2013]: A multiclass image classification data set with 10 overlapping categories between the Office data set and Caltech256 data set. The off-the-shelf black-box classifier is trained on the office dataset. The Caltech images are used as the target domain.

Preprocessing: For textual features, words were converted to lower case and then stemmed. All the stop words, punctuation and symbols were removed. The tf-idf feature vector as a bag of words on n-grams $n = \{1, 2, 3, 4\}$ were extracted for each review. For the images, SIFT features for each image were calculated. Then each image is represented as a tf-idf feature vector on Bag of Visual Words (BoVW). The

BoVW are computed through K-Means on SIFT descriptors for each image. The number of clusters for convex non-convex data set is set to 100, and for cats and dogs data set the number of clusters is set to 600. The cluster size is chosen based on the 10-fold cross validation.

Table 4.7

Statistics of the Seven Data Sets

Data set	Type	# of examples
Binary data sets		
IMDB movie reviews	Text	10000
Amazon fine food reviews	Text	10000
Convex-nonconvex images	Image	3000
Cats and dogs images	Image	3000
News articles	Text	1395
Multiclass data sets		
Amazon product reviews	Multiclass Text	3521
Office-Caltech data set	Multiclass Image	2533

Comparison methods: The effectiveness of the proposed framework is demonstrated by comparing with the baseline off-the-shelf classifier and the strong baseline of SVM. The various methods compared in the experiments and their setup are as follows:

1. **FastDAM**: Fast Domain Adaptation Machine [Duan et al., 2009]. To compare with the proposed **AOT** framework, only one set of classification labels from the off-the-shelf classifier were considered for FastDAM.
2. **SVM**: Strong baseline SVM trained on the known labeled examples from the target domain.
3. **OTSC**: Off-the-shelf classifier. The Stanford sentiment classification toolkit is used as the off-the-shelf classifier for the binary sentiment classification data sets. For the image data sets, a logistic regression model trained on the similar images as the target domain is used as the off-the-shelf classifier.

Effectiveness of AOT : Effectiveness of the proposed **AOT** framework is evaluated by comparing with other methods. For all the experiments, the regularization parameters in the label deficiency **ADDRESSLABDEF** are set to $\mu_1 = 0.7$ and $\mu_2 = 0.5$. For all the experiments, the results are reported after 30 different runs on randomly sampled data from the training set. The effectiveness of the proposed **AOT** approach is evaluated from a sample of 10-140 labeled examples for the binary data sets, 20-280 examples for the multi-class data sets.

Figures 4.7–4.11 show the results on effectiveness for binary data sets. Figures 4.12 and 4.13 show the effectiveness for multi-class data sets.

From the results in the Figure 4.7, the proposed **AOT** framework performs better than all the competitors on both the text and image data sets. Its performance is

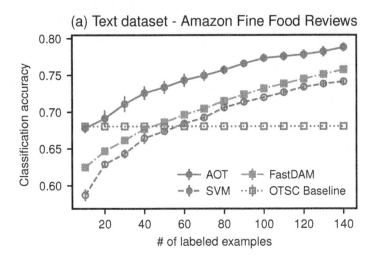

Figure 4.7 Classification accuracy on binary Amazon fine foods text dataset with 10-140 labeled examples.

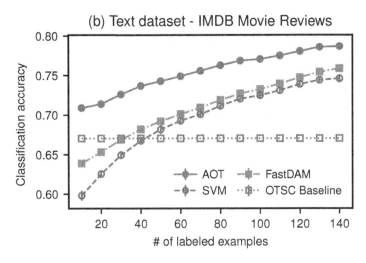

Figure 4.8 Classification accuracy on binary IMDB text data set with 10-140 labeled examples.

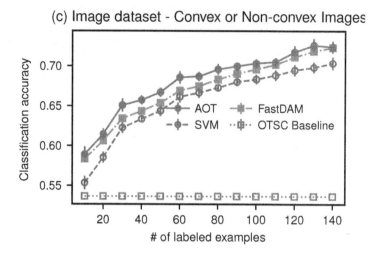

Figure 4.9 Classification accuracy on binary convex and non-convex images data set with 10-140 labeled examples.

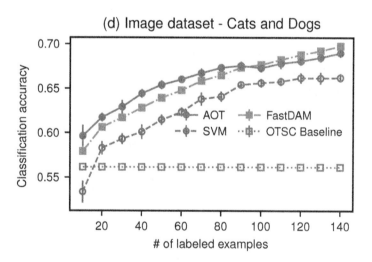

Figure 4.10 Classification accuracy on binary cats and dogs image dataset with 10-140 labeled examples.

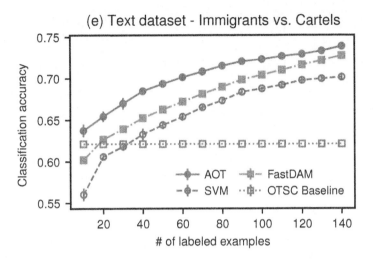

Figure 4.11   Classification accuracy on binary news articles text dataset with 10-140 labeled examples.

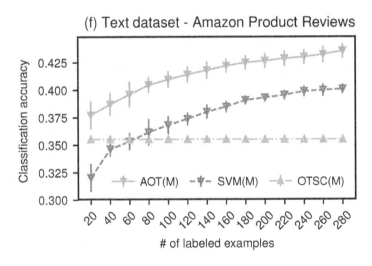

Figure 4.12   Classification accuracy on multi-class Amazon product reviews text dataset with 20-280 labeled examples.

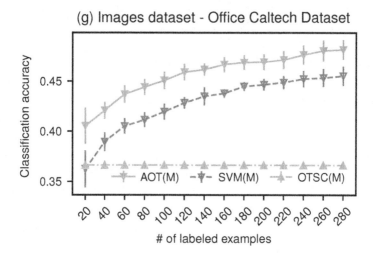

Figure 4.13 Classification accuracy on multi-class Office-Caltech image dataset with 20-280 labeled examples.

very close to that of FastDAM on both image data sets. This is because the labels generated by the baseline off-the-shelf classifier are very noisy on the image data sets, and the gain achieved by the proposed **AOT** framework is limited by the quality of the labels generated by the off-the-shelf classifier.

4.3.3.1 Two Stage Analysis

We analyze the benefit of addressing label deficiency and distribution shift individually. The number of labeled examples is set to $m = 140$. We evaluate the performance of algorithms ADDRESSLABDEF, ADDRESSDISSHIFT and **AOT** on *Amazon fine food reviews* and *Cats and dogs* data sets.

From Figures 4.14 and 4.15, it can be observed that addressing label deficiency through manifold regularization alone is more helpful than addressing distribution shift. Also combining both algorithms performs better than the performance of the individual algorithms. This demonstrates the power of combining both algorithms together for better adaptation results.

4.3.3.2 Sensitivity Analysis

In this subsubsection, we analyze the influence of hyper-parameters on the proposed **AOT** framework. We analyze the influence of hyper-parameters through grid search as shown in Figure 4.16. Both parameters μ_1 and μ_2 in the objective function taking values in the interval $[0, 1]$ are analyzed. In general, the performance of the proposed framework is robust to small perturbations in the parameters. Furthermore, it was observed that the parameter μ_1 which controls the influence of the regularizer on the labeled examples gives good results with higher values $\mu_1 \geq 0.6$ and performs poorly

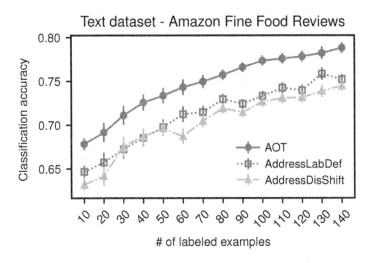

Figure 4.14 Two stage analysis for the *Amazon fine foods* binary text data set.

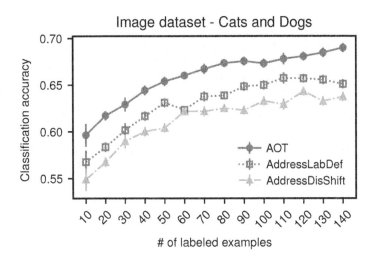

Figure 4.15 Two stage analysis for the *cats and dogs* binary image data set.

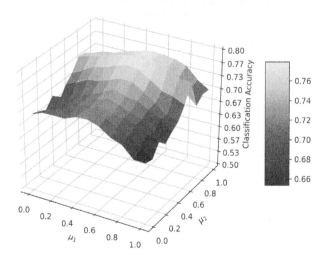

Figure 4.16 Sensitivity analysis of the AOT on IMDB dataset.

with smaller values. Also, the parameter μ_2 which controls the sparsity has a better accuracy for a balance elastic net regularizer around $\mu_2 = 0.5$.

4.3.3.3 Convergence Analysis

In this subsection, we analyze the convergence of the proposed AOT framework. Figure 4.17 shows the convergence of the cost function from eq. (4.25) on IMDB data set. The number of iterations for the algorithms ADDRESSLABDEF and ADDRESS-DISSHIFT is set to $k = 10$. It can be seen that the algorithm converges around $T = 60$ iterations.

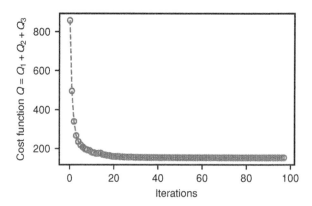

Figure 4.17 Convergence of the AOT on IMDB dataset.

4.3.3.4 Runtime Analysis

Figure 4.18 shows the runtime in seconds for varying sizes of the data set.

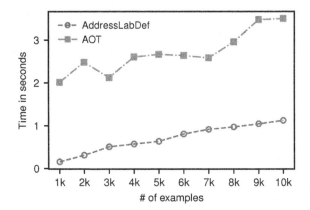

Figure 4.18 Running time of the ADDRESSLABDEF and AOT algorithms on IMDB data set.

The number of labeled examples is set to $m = 140$ for all the cases, the remaining examples are considered as unlabeled. In our experiments, the **AOT** algorithm converges after $T = 60$ iterations. For ADDRESSLABDEF we have set the max number of iterations $K = 100$.

5 Explainable Transfer Learning

Traditional machine learning typically requires a large amount of quality labeled data to obtain reliable predictive models. Finding such quality labeled data can be expensive and time-consuming, sometimes requiring inputs from domain experts, such as health imagery like MRI scans [Liu et al., 2018] and fine-grained classification [Zhao et al., 2017]. This has motivated the research on *transfer learning* [Pan and Yang, 2010], which leverages the knowledge and insights obtained from a source domain with plenty of labeled data to help build predictive models in a target domain with limited or even no labeled data at all. The major challenge here is how to model the relatedness between the source and target domains despite the different data distributions. Up until now, various transfer learning techniques have been proposed and successfully applied to a wide variety of real-world problems such as machine translation [Wu et al., 2008], image classification [Long et al., 2016], web document classification [Weiss et al., 2016], etc. With the recent advances in deep learning techniques, a variety of deep neural network models have been proposed for transfer learning, often leading to significant improvement in the performance. [Hoffman et al., 2018] showed that knowledge is transferable from synthetic to real-world domains, Long et al. [2015] proposed a deep adaptation network that can learn transferable features from pre-trained models, popularly known as fine-tuning, Young et al. [2018] observed that deep learning based approaches were effective in natural language processing tasks.

On the other hand, with increasingly complex predictive models, it becomes more and more challenging to explain the model outputs to end users in such a way that can be comprehended by non-experts of data mining. To address this challenge, quite a few efforts have been made to explain model outputs in recent years. For example, Ribeiro et al. proposed an explanation framework to explain individual predictions of black box models; in [Lundberg and Lee, 2017], the authors proposed a unified approach connecting game theory with local explanations; Koh and Liang [2017] proposed a framework based on influence functions that helps in efficiently computing the influence of training examples on test data; Selvaraju et al. [2017] showed that the activation weights in convolution neural networks could be learned using the gradients on the activation maps without retraining the model. However, in the context of transfer learning, due to the difference in data distribution between the source and target domains, existing explanation techniques cannot be readily applied to interpret the transfer learning models. In other words, the key challenge here is how to provide the explanation for the predictive model in the *target domain* using the information obtained from the *source domain* (e.g., the influence of examples, relevance of features). In particular, some source domain examples might contribute more to the predictive model in the target domain, whereas others might be irrelevant or even cause the negative transfer [Pan and Yang, 2010]. Similarly, some features may provide more meaningful information, or possess more discriminative power compared to other background or unnecessary features [Blitzer et al., 2006, Weiss et al., 2016]. A good explanation model should capture the information on important examples and relevant features to provide coherent explanations of the model behavior [Doshi-Velez and Kim, 2017]. Motivated by this, we learn the importance weights, and the set relevant features associated with the source domain examples and use them to interpret the behavior of the transfer learning model.

In this chapter, we propose an **explainable Transfer Learning** framework (*exTL*) that learns importance weights from the source domain examples and identifies the relevant set of features that are conducive to transfer. In our approach we sample a mini-batch of examples from the source and target domains, perturbing the source domain examples by a small weight and estimate its influence through performance on the set of labeled examples in the target domain. The key idea here is that those examples that are similar to the examples in the target domain need to be up-weighted. The optimal weights are computed in an online fashion, and the model parameters are updated based on the reweighted cost function. We also enforce a regularizer that helps in learning a domain invariant feature space which further enables transfer from the source domain to the target domain. Using only a small set of labeled examples in the target domain can lead to model overfitting; we prevent this by considering a set of unlabeled examples in the mini-batch with low entropy. This set of unlabeled examples helps in learning a more generalizable model to the target domain and also in learning the domain invariant space between the source and target domains. To be specific, the novel contributions of the work on explainable transfer learning are as follows:

1. *exTL* : A novel semi-supervised transfer learning framework that learns the importance weights in an online fashion along with a set of relevant features conducive to transfer and a domain invariant space spanning the source and the target domains.
2. An algorithm that applies to shallow and deep neural network models for explainable transfer learning. Theoretical analysis with the shallow network example and exhaustive empirical analysis on two text and one image data set to demonstrate the performance of the proposed approach on multiple baselines and widely used transfer learning approaches.
3. A novel approach to using the importance weights on the source domain examples and the features to interpret the underlying transfer learning model. We provide analysis on the reliability of the model and demonstrate the superiority of our approach through visualization of important features in the image data set.

The rest of the chapter is organized as follows. We discuss the proposed framework along with the algorithm and theoretical analysis in Section 5.1. In Section 5.2, we perform an empirical analysis of two text data sets and an image data set.

5.1 PROPOSED APPROACH

In this section, we propose our novel approach *exTL* to learn importance weights and relevant features that foster knowledge transfer from the source to the target domain. We start by introducing the notation, the *exTL* framework, online algorithm to compute weights and relevant features and finally discuss our set up with an example using a shallow neural network.

5.1.1 NOTATION

Let S denote the source domain and T denote the target domain. We consider that the labeled source data is abundant and the labeled data in the target domain is scarce. Let $\mathcal{D}_S = \{\mathcal{X}_S, \mathcal{Y}_S\}$ represent the examples from the source domain, i.e., and let $\mathcal{X}_S = \{\mathbf{x}_1^S, \ldots, \mathbf{x}_m^S\} \subset \mathbb{R}^d$ be the set of m labeled examples from the source domain, and d be the dimensionality of the feature space. Let $\mathcal{Y}^S = \{y_1^S, \ldots, y_n^S\} \subset 0, 1^{m \times |C|}$, where $|C|$ is cardinality of set of classes $\{1, \ldots, C\}$. For convenience we represent features in the shared feature space of size d. Similarly, let $\mathcal{D}_T = \{\mathcal{X}_T^l, \mathcal{Y}_T^l\} \cup \{\mathcal{X}_T^u\}$ represent the union of a small set of labeled examples and a large set of unlabeled examples from the target domain. Let $\mathcal{X}_T^l = \{\mathbf{x}_1^T, \ldots, \mathbf{x}_{n_l}^T\} \subset \mathbb{R}^d$ be the set of n_l labeled examples from the target domain, and $\mathcal{X}_T^u = \{\mathbf{x}_1^T, \ldots, \mathbf{x}_{n_u}^T\} \subset \mathbb{R}^d$ be the set of n_u unlabeled examples from the target domain. Let $\mathcal{P}_S(\mathcal{X}_{\mathcal{S}})$ and $\mathcal{P}_T(\mathcal{X}_{\mathcal{T}})$ characterize the probability distributions of the source and the target domains. In this paper we consider a transfer learning setup where probability distributions $\mathcal{P}_S \neq \mathcal{P}_T$.

5.1.2 EXTL FRAMEWORK

Let $f(x, \theta)$ be neural network model a shallow or deep network with model parameters θ and $\mathcal{L}(y, f(x, \theta))$ be the loss function, for example entropy loss for multiclass settings. The goal of the transfer learning is to learn the model that can improve the classification accuracy on the unlabeled examples in the target domain. The objective function for empirical risk minimizer that minimizes the loss on the known labeled examples is given as follows:

$$\theta^* = \underset{\theta}{\arg\min} \sum_{i=1}^{m} \mathcal{L}(y_i^S, f(\mathbf{x}_i^S, \theta)) + \sum_{i=1}^{n_l} \mathcal{L}(y_i^T, f(\mathbf{x}_i^T, \theta)) \qquad (5.1)$$

Directly training the classifier on labeled examples may not perform well if a meaningful representation across the domains is not established; it can even lead to overfitting of the model to data in the source domain. Such overfitting can result in reduced performance of the classifier on target domain data $\mathcal{D}_T$. Our intuition is that if we can reweight the source domain examples and also simultaneously learn domain invariant feature space by minimizing the distribution shift in between the source and target domains, it is possible to train the classifier on source domain data $\mathcal{D}_S$ that performs well on the target domain $\mathcal{D}_T$. With that intuition, we propose our novel transfer learning approach that reweights source domain examples and at the same time learns the domain invariant representation across the source and target domains.

The loss function $\mathscr{L}$, parameterized by θ and $\mathbf{w}$ for our approach is given as follows:

$$\mathscr{L}(\theta, \mathbf{w}) = \sum_{i=1}^{m} \mathscr{L}(y_i^S, \mathbf{w}_i f(\mathbf{x}_i^S, \theta)) + \sum_{i=1}^{n_l} \mathscr{L}(y_i^T, f(\mathbf{x}_i^T, \theta)) \\ + \lambda d^2(\mathscr{X}_S, \mathscr{X}_T) \tag{5.2}$$

where $\mathbf{w}_i$ are the weights for reweighting source domain examples, and the second term $d(\mathscr{X}_S, \mathscr{X}_T)$ represents the distance between source domain data $\mathscr{X}_S$ and target domain data $\mathscr{X}_T$ with the penalty hyperparameter $\lambda > 0$. More details regarding importance weight computation, domain invariant space and relevant feature identification are discussed in the following subsections.

5.1.3 REWEIGHTING THE SOURCE DOMAIN EXAMPLES

In most practical scenarios, not all the examples can be considered equally representative regarding the target domain. There could be a set of examples with data distribution similar to the target domain which have a positive influence for knowledge transfer, and examples that are very dissimilar can lead to negative transfer. Identifying and uplifting the contribution of such similar examples will often lead to better model performance [Shimodaira, 2000, Pan and Yang, 2010]. The idea is to learn the weights on such similar examples and let them guide the model parameter learning procedure.

Given the source domain data, we aim to minimize the expected loss by reweighting examples of the source domain. Let $\mathbf{w}_i$ be the weight given to the example from the source domain; the objective function that learns the optimal model θ^* with input reweighting is given as follows:

$$\theta^*(\mathbf{w}) = \underset{\theta}{\arg\min} \sum_{i=1}^{m} \mathscr{L}(y_i^S, \mathbf{w}_i f(\mathbf{x}_i^S, \theta)) \tag{5.3}$$

In our approach the weights $\{\mathbf{w}_i\}_{i=1}^{m}$ are considered to reweight the loss contribution of each of the source domain examples. Unlike classical input reweighting approaches [Shimodaira, 2000] we treat the weights as hyperparameters that drive in training a better transfer learning model. Given the set of labeled examples from the target domain, a straight forward way for optimal selection of the weights w is based on the performance of the classifier on the known labeled examples [Ren et al., 2018]. The optimal selection of the weights w^* is given as:

$$\mathbf{w}^* = \underset{\mathbf{w}, \mathbf{w} \geq 0}{\arg\min} \sum_{i=1}^{n_l} \mathscr{L}(y_i^T, f(\mathbf{x}_i^T, \theta^*(\mathbf{w}))) \tag{5.4}$$

Here we consider the weights to be positive, since minimizing the negative training loss can result in an unstable behavior. Considering only a small set of labeled examples for optimal weight selection has two major issues: (1) the number of labeled examples in the target domain is considered to be very small, therefore it is difficult to learn optimal weights from such a small set of examples; (2) trying to fit the weights to the small size of labeled examples can result in overfitting of the model to the labeled set and therefore not generalizable to the target domain. To avoid such issues, we propose to leverage the larger set of unlabeled examples, posing our approach *exTL* as a semi-supervised learning problem. In our approach, we sample a small set of unlabeled examples, compute the noisy labels for the model and use that subset of examples with low entropy. For a cross-entropy loss function, low loss examples can be considered. The authors in [Han et al., 2018] showed that by using the low loss examples, it is possible to learn a robust classifier. Let n'_u represent the small subset of examples with classification loss less than a threshold τ. The parameter *tau* can be set by performance of the model on known labeled examples from the target domain. The objective function to compute the optimal weights is

$$\mathbf{w}^* = \underset{w,w \geq 0}{\operatorname{argmin}} \sum_{i=1}^{n_l} \mathcal{L}(y_i^T, f(\mathbf{x}_i^T, \theta^*(\mathbf{w}))) + \sum_{i=1}^{n'_u} \mathcal{L}(y_i^T, f(\mathbf{x}_i^T, \theta^*(\mathbf{w}))) \tag{5.5}$$

It can be seen from eq. (5.3) and eq. (5.5) that calculating the optimal weight $\mathbf{w}_i$ requires two nested loop operations which is very expensive to compute and hard to scale to large data sets. Instead, we want to follow recent work on estimating the influence of training examples on test data through perturbation of training examples with small weight ε. As shown in [Koh and Liang, 2017], the goal here is to understand the influence of source domain examples towards the performance on the small labeled data set from the target domain. We would give higher weights to the examples that have more influence which are usually examples similar to the target domain examples. We consider by perturbing each training example with a small weight ε, let the reweighted version of the loss function from eq. (5.5) be $\mathcal{L}_{i,\varepsilon}(\theta) = \varepsilon \mathcal{L}_i(\theta)$; then the optimal ε^* can be computed by minimizing the loss on the set of labeled examples from the target domain.

$$\varepsilon_t^* = \underset{\varepsilon}{\operatorname{argmin}} \frac{1}{n_l} \sum_{i=1}^{n_l} \mathcal{L}_{i,\varepsilon}(\theta_{t+1}(\mathbf{w})) + \frac{1}{m'_u} \sum_{i=1}^{m'_u} \mathcal{L}_{i,\varepsilon}(\theta_{t+1}(\mathbf{w})) \tag{5.6}$$

Considering optimization of the model through Stochastic Gradient Descent (SGD), computing the ε^* in eq. (5.6) for each mini-batch of examples is expensive, so instead we take a single gradient descent step on a mini-batch of labeled examples from the target domain. For each model training iteration, we inspect the descent direction of a mini-batch of examples from the source domain locally on the training loss surface and reweight them according to their similarity to the descent direction of the loss surface computed on the set of labeled target domain examples. At every step of training the model, a mini-batch of labeled examples from the source domain is sampled and the parameters of the model θ are adjusted according to the descent direction of the expected loss on the mini-batch. The gradients are rectified to get a non-negative weighting. Let b_T^l be the mini-batch of labeled examples from the target domain and b_T' be low entropy examples from the mini-batch of b_T^u unlabeled examples.

The weight computation step with $(b_T^l + b_T')$ mini-batch examples at iteration t is given as follows:

$$w_{i,t} = -\eta \frac{\partial}{\partial \varepsilon_{i,t}} \frac{1}{(b_T^l + b_T')} \sum_{i=1}^{(b_T^l + b_T')} \mathcal{L}(y_i^T, f(\mathbf{x}_i^T, \theta_{t+1}(w))) \tag{5.7}$$

$$\tilde{w}_{i,t} = \max(w_{i,t}, 0) \tag{5.8}$$

where η is the descent step size on ε. Further, to keep the weights comparable across the mini-batches, we normalize the weights, to sum up to one for each mini-batch.

5.1.4   DOMAIN INVARIANT REPRESENTATION

As the data distribution is different between the source and target domains, for an effective transfer learning it is important to learn a domain invariant feature space. To minimize the distance between the data distributions, we employ Maximum Mean Discrepancy (MMD) [Gretton et al., 2012, Tzeng et al., 2014]. The distance between the examples in the source domain and the target domain is computed with the representation $\phi(\cdot)$. The representation $\phi(\cdot)$ can be considered as an input feature mapping, for example word embeddings in the case of text, outputs of hidden layers in case of deep neural networks. Representation $\phi(\cdot)$ on the source domain examples $\mathcal{X}_S$ and target domain examples $\mathcal{X}_T$. The empirical approximation of the distance d_k for the kernel k is given by:

$$d_k^2 = \left\| \frac{1}{|\mathcal{X}_S|} \sum_{\mathbf{x}_i^S \in \mathcal{X}_S} \phi(\mathbf{x}_i^S) - \frac{1}{|\mathcal{X}_T|} \sum_{\mathbf{x}_i^T \in \mathcal{X}_T} \phi(\mathbf{x}_i^T) \right\|^2 \tag{5.9}$$

From eq. (5.9) it can be seen that the MMD distance d_k between the source and the target domains is defined as the distance between the mean embeddings. MMD was earlier studied towards *two-sample testing*, where acceptance or rejection decisions are made for a null hypothesis $\mathcal{P}_S = \mathcal{P}_T$, the two samples belong to the same distribution. In [Gretton et al., 2012], it was shown that the distributions $\mathcal{P}_S = \mathcal{P}_T$, if and only if the $d_k^2 = 0$. Here k is the characteristic kernel associated with the feature map ϕ, given as $k(\mathcal{X}_S, \mathcal{X}_T) = <\phi(\mathcal{X}_S), \phi(\mathcal{X}_T)>$. It was theoretically proved that the kernel for the mean embedding of the two distributions is critical for the test power and low test error [Tzeng et al., 2014, Long et al., 2015]. Empirically the estimate of MMD computes the square distance between the empirical kernel mean embeddings; expanding eq. (5.9) and applying the definition of kernel function k, we have:

$$\hat{d}_k^2 = \frac{1}{m^2} \sum_{i=1}^{m} \sum_{j=1}^{m} k(\mathbf{x}_i^S, \mathbf{x}_j^S) + \frac{1}{n^2} \sum_{i=1}^{n} \sum_{j=1}^{n} k(\mathbf{x}_i^T, \mathbf{x}_j^T)$$
$$- \frac{2}{mn} \sum_{i=1}^{m} \sum_{j=1}^{n} k(\mathbf{x}_i^S, \mathbf{x}_j^T) \tag{5.10}$$

where k is the kernel and $\hat{d}_k$ is the unbiased estimator of d_k. In practice, a kernel is chosen based on the model and data set. For example, a distance metric like cosine

---

Algorithm 6: *exTL* Algorithm

1: **Input:** Source and target domain data sets $\mathscr{D}_S$ and $\mathscr{D}_T$, mini-batch size for sampling source and target domain examples b_S, b_T^l and b_T^u, initial model parameters θ_0 and max number of iterations T. Hyperparameters τ, α, β and η.

2: **Initialization:**

3: Set $t = 0$.

4: **Repeat:**

5:    (i). Sample the mini-batch examples from $\mathscr{D}_S$ and $\mathscr{D}_T$.

$$\{\mathscr{X}_S^{\cdot b_S}, \mathscr{Y}_S^{\cdot b_S}\} \leftarrow \text{MINI-BATCH}(\mathscr{D}_S, b_S)$$

$$\{\mathscr{X}_T^{\cdot b_T^l}, \mathscr{Y}_T^{\cdot b_T^l}\} \leftarrow \text{MINI-BATCH}(\mathscr{D}_T^l, b_{TL})$$

$$\{\mathscr{X}_T^{\cdot b_T^u}\} \leftarrow \text{MINI-BATCH}(\mathscr{D}_T^l, b_{TU})$$

6:    (ii). Compute $\hat{\theta}_t$ based on $\mathscr{D}_S$.

$$\hat{\mathscr{Y}}_S^{b_S} \leftarrow forward(\mathscr{X}_S^{\cdot b_S}, \theta_t); \varepsilon \leftarrow 0;$$

$$\mathscr{L}_S \leftarrow \mathscr{L}(\hat{\mathscr{Y}}_S^{b_S}, \mathscr{Y}_S^{b_S}, \varepsilon, \theta_t); \hat{\theta}_t \leftarrow \theta_t - \alpha \nabla \mathscr{L}_S$$

7:    (iii). Update the parameters ε based on $\mathscr{D}_T$.

$$\hat{\mathscr{Y}}_T^{b_T^l} \leftarrow forward(\mathscr{X}_T^{\cdot b_T^l}, \theta_t); \hat{\mathscr{Y}}_T^{b_T^u} \leftarrow forward(\mathscr{X}_T^{\cdot b_T^u}, \theta_t);$$

$$b' \leftarrow H(\hat{\mathscr{Y}}_T^{b_T^u}) < \tau; \hat{\mathscr{Y}}_T^{b_T} \leftarrow \{\hat{\mathscr{Y}}_T^{b_T^l}, \hat{\mathscr{Y}}_T^{b'}\}$$

$$\mathscr{L}_T \leftarrow \mathscr{L}(\hat{\mathscr{Y}}_T^{b_T}, \mathscr{Y}_T^{b_T}, \varepsilon); \nabla \varepsilon \leftarrow \nabla \mathscr{L}_T$$

$$\tilde{w} \leftarrow \max(-\nabla \varepsilon, 0); w \leftarrow normalize(\tilde{w})$$

8:    (iv). Reweight the loss for each of the source domain examples using w and update the parameters θ.

$$\mathscr{L} \leftarrow \mathscr{L}(\hat{\mathscr{Y}}_S^{b_S}, \mathscr{Y}_S^{b_S}, w, \theta_t); \theta_{t+1} \leftarrow \theta_t - \beta \nabla \mathscr{L}$$

9:    (iv). Every p iterations, compute the feature relevance scores using DeepSHAP and prune the feature by zeroing out their input contributions in further iterations.

10: **While** $t < T - 1$

11: **Output:** The adapted parameters for the model θ_t

---

similarity can be used for linear models and text data sets, whereas for an image data set a Gaussian kernel, $k(\mathbf{x}_i, \mathbf{x}_j) = exp^{-||\mathbf{x}_i - \mathbf{x}_j||^2 / \gamma}$ with bandwidth γ set to the median pairwise distances on the data [Gretton et al., 2012].

Furthermore, we compute the feature relevance scores using the Deep SHAP [Lundberg and Lee, 2017] approach based on Shapely Values and Deep LIFT [Shrikumar et al., 2017]. Deep SHAP through back propagation, combines SHAP values computed for various neurons in the network into SHAP values for the whole network. In our approach, we estimate feature relevance scores of the network every p iterations and remove the set of features with the least contribution. The goal of pruning the features is to remove the background features that do not contribute substantially to the model. Also, having smaller features makes it easier to explain the underlying model [Ribeiro et al.].

5.1.5   ALGORITHM

In this section, we present our **exTL** algorithm along with the analysis of the convergence and upper bound on the target domain risk. Algo.6 illustrates the proposed algorithm. Let T be the number of maximum iterations. For simplicity, we initialize the model parameter to zeros. For each model training iteration, we sample a mini-batch b_S from the source domain, b_T^l labeled examples and b_T^u unlabeled examples from the target domain. Based on eq. (5.3), estimate $\hat{\theta}$ from the mini-batch of source domain examples. From the mini-batch samples from the target domain, filter the set of unlabeled examples with low entropy $(H < \tau)$. Combining the b_T^l labeled examples and b_T' filtered examples from the target domain, let $b_T = (b_T^l + b_T')$, take a gradient step on the loss function as shown in eq. (5.5). Estimate the weights as shown in eq. (5.7) and eq. (5.8). Using the weights estimated we update the model parameters along with the MMD loss computed over the mini-batch of examples from the source domain and target domain. It can be seen that the weights are computed on the mini-batch examples only; considering the random sampling scenario, we accrue the contributions for each example from the mini-batch and average it over all the iterations for global importance weight computation. The feature importance values computed in the last iteration are considered as global feature importance values. All the step-size hyperparameters α, β and η are set using the line-search.

We show that by using the labeled examples from the target domain, the loss function in the proposed algorithm is monotonically decreasing. As the loss function on the target domain examples is monotonically decreasing over the iteration steps it can be considered that the model trained using the source domain examples learns a good model on the target domain. Also, our algorithm converges as shown in Theorem 5.1. The pruning of features through feature relevance scores eliminates noisy and irrelevant features which also helps in convergence and in most practical cases does not impact the model. We also provide the upper bound on the expected target risk of the distribution in Theorem 5.2.

Theorem 5.1

The risk function $\hat{\mathscr{R}}_T$ is monotonically decreasing for any sequence of training batch iterations.

$$\hat{\mathscr{R}}_T(\theta_{t+1}) \leq \hat{\mathscr{R}}_T(\theta_t) \tag{5.11}$$

where t is the training iteration.   ∎

Proof. Suppose the loss function is Lipschitz continuous with constant L and the loss function $\mathscr{L}(\theta)$ on the set of training examples has bounded gradients, $||\nabla\mathscr{L}(\theta)||^2 \leq \gamma, \forall \mathbf{x} \subset \mathbb{R}^d$. Let the learning rate at iteration step t given by α_t satisfy the condition

$\alpha_t \leq \frac{2bs}{L\gamma^2}$, where b is the batch size of the mini-batch. For the model f parametrized by θ, the risk on the target domain examples can be given as:

$$\mathscr{L}(\theta) = \frac{1}{n_l} \sum_{j=1}^{n_l} \mathscr{L}\left(f(\theta, \mathbf{x}_j, y_j)\right) \tag{5.12}$$

For every iteration of the algorithm, as shown in Section 3.2, we also sample a set of unlabeled examples from the target domain. The total number of examples considered for every mini-batch is given as $b_T = (b'_T + b^l_T)$. We choose b'_T such that the total loss incurred by the examples is less than τ. We set the τ to be less than the loss incurred by the set of labeled examples from the target domain. For each iteration t in the algorithm, the risk on the target domain examples for every mini-batch can be given as:

$$\begin{aligned} \mathscr{L}(\theta_t) &= \frac{1}{b^l_T} \sum_{j=1}^{b^l_T} \mathscr{L}\left(f(\theta, \mathbf{x}_j, y_j)\right) + \frac{1}{b^l_T} \sum_{j=1}^{b^l_T} \mathscr{L}\left(f(\theta, \mathbf{x}_j, y_j)\right) \\ &\leq \frac{2}{b^l_T} \sum_{j=1}^{b^l_T} \mathscr{L}\left(f(\theta, \mathbf{x}_j, y_j)\right) \end{aligned} \tag{5.13}$$

The risk is bounded by a constant factor multiplied with the risk $\hat{\mathscr{R}}_T(\theta)$, so it is safe to consider that the sampling from the unlabeled examples is similar to the risk on labeled examples.

Similarly, let empirical loss on the source domain examples is given as:

$$\mathscr{L}(\theta) = \frac{1}{m} \sum_{i=1}^{m} \mathscr{L}\left(f(\theta, \mathbf{x}_i, y_i)\right) \tag{5.14}$$

Following the computation similar to the shallow neural network example (Section 3.4), the update rule for computing the model parameters can be given as:

$$\theta_{t+1} = \theta_t - \frac{\alpha_t}{m} \sum_{i=1}^{b_S} \max\left(\nabla \mathscr{L}_j^\top \nabla \mathscr{L}_i, 0\right) \mathscr{L}_i \tag{5.15}$$

which can be rewritten as:

$$\begin{aligned} \theta_{t+1} - \theta_t &= \frac{\alpha_t}{m} \sum_{i=1}^{b_S} \max\left(\nabla \mathscr{L}_j^\top \nabla \mathscr{L}_i, 0\right) \mathscr{L}_i \\ &= \Delta\theta \end{aligned} \tag{5.16}$$

where α_t is the learning rate at time-step t.

Since the loss function is lipschitz-smooth, and considering that $\max\left(\nabla\mathscr{L}_j^\top\nabla\mathscr{L}_i,0\right)\nabla\mathscr{L}_j^\top\nabla\mathscr{L}_i \leq \max\left(\nabla\mathscr{L}_j^\top\nabla\mathscr{L}_i,0\right)^2$, we have:

$$\mathscr{L}(\theta_{t+1}) \leq \mathscr{L}(\theta_t) + \nabla\mathscr{L}_j^\top\Delta\theta + \frac{L}{2}||\Delta\theta||^2$$

$$\leq \mathscr{L}(\theta_t) - \frac{\alpha_t}{m}\sum_{i=1}^{b_S}\max\left(\nabla\mathscr{L}_j^\top\nabla\mathscr{L}_i,0\right)^2 \qquad (5.17)$$

$$+\frac{L}{2}\frac{\alpha_t^2}{m^2}\sum_{i=1}^{b_S}\max\left(\nabla\mathscr{L}_j^\top\nabla\mathscr{L}_i,0\right)^2$$

If we denote $\Omega_t = \sum_{i=1}^{b_S}\max\left(\nabla\mathscr{L}_j^\top\nabla\mathscr{L}_i,0\right)^2$, we can express the loss at iteration $t+1$ is given as:

$$\mathscr{L}(\theta_{t+1}) \leq \mathscr{L}(\theta_t) - \frac{\alpha_t}{m}\Omega_t\left(1 - \frac{L\alpha_t\sigma^2}{2n}\right) \qquad (5.18)$$

as Ω_t is non-negative and as $\alpha_t \leq \frac{2n}{L\sigma^2}$ the target loss converges. $\qquad\square$

Theorem 5.2

Let $\theta \in \mathscr{H}$ be a hypothesis, $\mathscr{R}_S(\theta)$ and $\mathscr{R}_T(\theta)$ be the expected risk of the source and target domains respectively, then

$$\mathscr{R}_T \leq \mathscr{R}_S + 2d_k(\mathscr{P}_S,\mathscr{P}_T) + C \qquad (5.19)$$

where C is a constant for the complexity of the hypotheis space $\mathscr{H}$ and the risk of an ideal hypothesis for the source and the target domains. $\qquad\blacksquare$

Proof. We present a proof sketch for the upper bound of the estimated target risk. Following the work from Ben-David et al. [2010], it shows that the $\mathscr{R}_T \leq \mathscr{R}_S + 2d_{\mathscr{H}}(\mathscr{P}_S,\mathscr{P}_T) + C_0$, where $d_{\mathscr{H}}(\mathscr{P}_S,\mathscr{P}_T)$ is the $\mathscr{H}$-divergence between the probability distributions $\mathscr{P}_S$ and $\mathscr{P}_T$. Following a similar approach to Long et al. [2016] it can be proved that by using the MMD based approach the $\mathscr{H}$-divergence between the domains can be reduced and thereby the estimated target risk can have an upper bound. $\qquad\square$

5.1.6 SHALLOW NEURAL NETWORK: AN EXAMPLE

In this example, we show how to compute weights vector **w** on the source domain examples for a shallow neural network model. The crucial step to computing the weights is through computing the gradients of the loss on the labeled target domain examples for small perturbation ε. Let d be the input size, h be the number of neurons in the hidden layer and let there be a single output neuron in the output layer. Let us consider a standard neural network (NN) architecture with a single hidden layer. For simplicity, consider a binary classification set up; the loss function $\mathscr{L}$ is considered to be the logistic loss. Let $\mathbf{x} \subset \mathbb{R}^d$ represent the input example, y be the true label for the input $\mathbf{x}$ and $\hat{y}_t$ be an estimated output from the neural network after t iterations for the input $\mathbf{x}$. The hidden layer learns a function $f_h : \mathbf{x} \to \mathbb{R}^h$ that maps the given input to the h-dimensional representation and is parametrized by $\mathbf{W}_1 \subset \mathbb{R}^{h \times d}$ and bias $c_1 \subset \mathbb{R}^h$. For an activation function σ, the function f_h is given as:

$$f_h(\mathbf{x}; \mathbf{W}_1, c_1) = \sigma(\mathbf{W}_1 \mathbf{x} + c_1) \tag{5.20}$$

Similarly, the output layer learns the function $f_o : \mathbb{R}^h \to [0,1]$ that is parametrized by a $\mathbf{W}_2 \subset \mathbb{R}^{1 \times h}$ and bias $c_2 \in \mathbb{R}$. Let the output from the hidden layer be represented by $\mathbf{z} \subset \mathbb{R}^h$, the function f_o be given as:

$$f_o(\mathbf{z}; \mathbf{W}_2, c_2) = \mathbf{W}_2 \mathbf{z} + c_2 \tag{5.21}$$

For the logisitic loss function $\mathscr{L}_{log}$, the loss w.r.t the output f_o is $\mathscr{L}_{log} = \log(1 + \exp(-y f_o))$. To estimate the optimal ε^* as shown in eq. (5.6) we can compute the gradients towards ε for the iteration t and parameters $\mathbf{W} = \{\mathbf{W}_1, \mathbf{W}_2\}$ as follows:

$$
\begin{aligned}
&\frac{\partial}{\partial \varepsilon_{i,t}} \mathbb{E}\left[\mathscr{L}(\mathbf{W}^{(t+1)}(\varepsilon))\Big|_{\varepsilon_{i,t}=0}\right] \\
&= \frac{1}{b'} \sum_{j=1}^{b'} \frac{\partial}{\partial \varepsilon_{i,t}} \mathscr{L}_j(\mathbf{W}^{(t+1)}(\varepsilon))\Big|_{\varepsilon_{i,t}=0} \\
&= \frac{1}{b'} \sum_{j=1}^{b'} \frac{\partial \mathscr{L}_j(\mathbf{W})}{\partial \mathbf{W}}\bigg|_{\mathbf{W}=\mathbf{W}_t}^{\top} \frac{\partial \mathbf{W}^{(t+1)}(\varepsilon_{i,t})}{\partial \varepsilon_{i,t}}\bigg|_{\varepsilon_{i,t}=0} \\
&\propto -\frac{1}{b'} \sum_{j=1}^{b'} \frac{\partial \mathscr{L}_j(\mathbf{W})}{\partial \mathbf{W}}\bigg|_{\mathbf{W}=\mathbf{W}_t}^{\top} \frac{\partial \mathscr{L}_i(\mathbf{W})}{\partial \mathbf{W}}\bigg|_{\mathbf{W}=\mathbf{W}_t} \\
&= -\frac{1}{m} \sum_{j=1}^{b'} \frac{\exp(-y.(\mathbf{W}_2 \mathbf{z}_j + c_2)) - y \mathbf{z}_j \nabla \mathbf{z}_j}{1 + \exp(-y.(\mathbf{W}_2 \mathbf{z}_j + c_2))} \\
&\quad \times \frac{\exp(-y.(\mathbf{W}_2 \mathbf{z}_j + c_2)) - y \mathbf{z}_j \nabla \mathbf{z}_j}{1 + \exp(-y.(\mathbf{W}_2 \mathbf{z}_j + c_2))} + C_3 d_k(z_i, z_j) \\
&= -\frac{1}{m} \sum_{j=1}^{b'} \left(C_1 \mathbf{z}_j \nabla \mathbf{z}_j\right)\left(C_2 \mathbf{z}_i \nabla \mathbf{z}_i\right) + C_3 d_k(z_i, z_j) \\
&= -\frac{1}{m} \sum_{j=1}^{b'} C_1 C_2 \left(\mathbf{z}_j^{\top} \mathbf{z}_i\right)\left(\nabla \mathbf{z}_j^{\top} \nabla \mathbf{z}_i\right) + C_3 d_k(z_i, z_j)
\end{aligned}
\tag{5.22}
$$

where C_1, C_2 and C_3 are some constants, j represents examples from the target domain and i represents the example from the source domain for which optimal ε_i

is being computed. It can be seen from the equation that the ε gradient depends on the dot product between the hidden inputs, the gradients of the hidden inputs and the distance between the inputs. For a shallow network, if the source domain input and target domain input are similar, the gradients are also similar and the distance between the source and target domain is small; then the corresponding source domain example is important and needs to be up-weighted. Otherwise, if they provide opposite gradients or the distance between the examples is large, they need to be down weighted.

5.2   RESULTS

In this section, we present our empirical analysis demonstrating the performance of our proposed *exTL* approach on both the text and image data sets. For each of the data sets we compare our approach with existing baselines and demonstrate its effectiveness; we also provide details regarding the reliability of the model along with visual explanations for image data set.

5.2.1   TEXT DATA

Data sets We analyze the performance on two binary classification data sets: (1) *Amazon product reviews* (AMA), the classic data set used for most transfer learning tasks [Pang and Lee, 2004]. The data set contains product reviews from *books*, *dvd*, *electronics* and *kitchen* domains. We considered a balanced set of 2000 reviews for each domain for a sentiment classification task. The bag-of-words feature vectors with tf-idf feature weighting are considered. We consider the top 5000 features with uni, bi, and tri-grams; (2) *Amazon, IMDB and YELP reviews* (AIY): User reviews for various products from Amazon [Pang and Lee, 2004], restaurant and business reviews from Yelp [Yelp, 2013] and movie reviews from IMDB [Maas et al., 2011b] are collected from different data sources. Each of the review ratings are normalized into sentiment classification labels positive or negative, and a binary classification data set is generated from the reviews. We consider a set of 10,000 examples for each of the domains. The bag-of-words feature vectors with tf-idf feature weighting and top 5000 features with uni, bi and tri-grams are considered for evaluation.

Effectiveness We analyze the effectiveness of our approach in transductive learning settings through the performance of the classifier for a sentiment classification task with a small set of labeled examples from the target domain. In the case of the AMA data set, we consider 100 randomly chosen balanced labeled examples from the target domain. For the AIY data set, we choose 200 balanced examples. As baselines, we consider Logistic Regression (LR) and SVM with Linear Kernel, transfer learning methods Subspace Alignment (SA) [Fernando et al., 2015] with dimensionality 100 and Transfer Component Analysis (TCA) [Pan et al., 2011] with 300 components. In our approach, we consider a shallow neural network with input layer the size of text data features and one hidden layer with size 100 and ReLU activation. We have used the linear kernel for the MMD distance regularizer. The model parameters are estimated by minimizing the logistic loss with SGD optimizer, parameterized with the learning rate set to 0.1 and drop the learning rate by half every 10 epochs. NN1 represents the baseline neural network model with one hidden layer. Our proposed method *exTL* is run for 40 epochs with 50 low contributing features pruned every epoch. To demonstrate the efficiency of this method, we also compare our approach with the one without feature pruning (*exTL*-AF, all features). The underlying model for *exTL* is similar to NN1.

In Table 5.1, we report classification accuracy for all the pairwise combinations of domains for the AMA data set. It can be seen that compared to the baselines our approach performs at par or better in most cases. The performance of the transfer learning approaches SA and TCA is worse than naive LR and SVM classifiers on an average. We presume this happens because only a small number of words are discriminative, and these words carry little weight in the sample transformation measures used. Moreover, both SA and TCA needs parameter tuning for effective performance. The proposed approach effectively converges and learns the set of important examples and relevant features with minimal to no parameter tuning.

In Table 5.2, we report the classification accuracy for all the pairwise combinations of domains for the AIY data set. Unlike the AMA data set, the AIY data set has a broader set of examples, and the domains and the underlying feature distributions

Table 5.1

Classification Accuracy on Amazon Product Reviews. Domains B:Books, D:Dvd, E:Electronics and K:Kitchen

SRC → TGT	B→D	B→E	B→K	D→E	D→K	E→K	D→B	E→B	K→B	E→D	K→D	K→E
LR	83.17	77.87	81.17	79.77	**81.77**	88.97	**80.97**	73.77	72.27	75.97	76.97	86.87
SVM	81.96	78.26	81.16	**79.86**	81.76	89.16	80.76	74.26	**73.86**	75.46	76.96	87.66
SA	58.74	62.94	63.04	59.84	67.14	69.04	65.04	62.94	58.74	60.34	61.84	71.14
TCA	58.82	63.22	66.32	61.62	64.12	74.02	58.12	52.02	57.52	57.42	60.12	70.52
NN1	83.38	**78.98**	81.48	79.38	81.48	88.08	73.18	73.88	72.88	76.18	**77.18**	**88.08**
exTL-AF	83.30	78.00	**81.60**	79.30	81.00	88.80	79.80	74.00	72.70	75.80	76.90	87.40
exTL	**83.40**	77.80	81.10	79.50	81.50	**89.40**	80.90	**74.70**	73.20	**76.20**	77.00	87.40

Table 5.2

Classification Accuracy on Amazon Product Reviews, Yelp and IMDB Movie Reviews. Domains A:Amazon, I:IMDB, Y:Yelp

SRC → TGT	A→I	A→Y	I→A	I→Y	Y→A	Y→I
LR	81.88	83.78	85.21	**84.98**	77.78	79.98
SVM	82.29	**83.89**	84.87	84.79	78.29	79.49
SA	66.91	63.81	71.11	69.01	66.91	64.31
TCA	67.25	65.65	68.15	62.15	56.05	61.45
NN1	81.80	83.36	85.46	77.16	77.86	80.16
exTL-AF	82.00	83.30	85.00	83.80	78.00	79.80
exTL	**82.96**	83.50	**85.79**	84.90	**79.21**	**80.43**

are more diverse. With a larger set of examples and different domains, our approach could learn better domain invariant representation, and parameter learning through importance weighting that results in better performance in most cases.

Reliability of the model We evaluate the reliability of the model following the set up similar to Ribeiro et al. We compare our approach to widely popular interpretable frameworks, LIME [Ribeiro et al.] and SHAP [Lundberg and Lee, 2017]. The reliability is measured based on self-interpretable models Sparse Logistic Regression (logistic regression with ll norm) and Decision Tree Classifier. We train both classifiers using all the examples from the source domain and the target domain. Training on the entire data set helps in learning the relevant set of discriminative features for the entire data set. For both classifiers, only the top 20 features are considered as the gold set for classifying training domain instances. The feature importance provided by LIME and SHAP are computed using a naive LR classifier as it provided good results in our effectiveness analysis on an average. For each of the unlabeled instances in the target domain, we compute the feature importance using the three methods and compute the fraction of the gold features that were recovered.

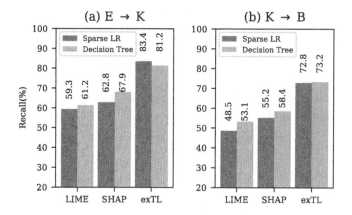

Figure 5.1   Recall on the important features.

Figure 5.1, shows the recall scores for two domain combinations for the AMA data set. In both the cases, our approach that learns the domain invariant space has a better recall compared to LIME and SHAP. Also, it can be observed that the two domain combinations Kitchen to Electronics has the best classification accuracy on average and Electronics to Books the least. In the case when the domains are more divergent, our approach works better in identifying the relevant features for transfer.

5.2.2 IMAGES

Data sets We analyze the performance of our approach on the **Office-Caltech** data set. The multi-class data set consists of images gathered from four different sources

Hoffman et al. [2013] with 10 different classes. A total of 1123 images are from the **Caltech10** data set, 956 samples from **Amazon** products, 157 samples from the **DSLR** camera and 295 samples from the **Webcam**.

Effectiveness Study We analyze the effectiveness of our approach in transductive learning settings through the performance of the classifier on a multi-class image classification task, with 10% labeled examples from the target domain. We consider the SURF based in order to compare with other baselines. SURF features are bag-of-visualwords histograms that are both scale and rotational invariant. Similar to the text dataset, we consider Logistic Regression (LR) and SVM with RBF Kernel, transfer learning methods Subspace Alignment (SA) with dimensionality 50 and Transfer Component Analysis (TCA) with 250 components and Gaussian kernel. Parameters for the baselines are selected through cross-validation on the labeled examples. For effectiveness analysis, we consider a simple shallow neural network with input layer the size of 800 and one hidden layer with size 200 and ReLU activation. We have used the Gaussian kernel with MMD regularizer. The model parameters are estimated by minimizing the cross-entropy loss with vanilla Stochastic Gradient Descent (SGD) optimizer, parameterized with the learning rate set to 0.1 and drop the learning rate by half every 5 epochs. NN1 represents the baseline neural network model with one hidden layer. Our proposed method *exTL* is run for 20 epochs with 10 low contributing features pruned every epoch. To demonstrate the efficiency of pruning, we also compare our approach without feature pruning with the model name (*exTL*-AF, all features). The underlying model for *exTL* is similar to a single hidden layer neural network model NN1.

Table 5.3

Classification Accuracy on Multi-class Classification of Office Caltech Dataset. Domains A:Amazon, C:Caltech10, D:Dslr, and W:Webcam

SRC → TGT	A → D	A → W	A → C	D → W	D → C	W → C	D → A	W → A	C → A	W → D	C → D	C → W
LR	0.382	0.325	0.447	0.688	0.288	0.302	0.281	0.305	0.477	0.809	0.384	0.275
SVM	0.401	0.312	0.443	0.688	0.256	0.279	0.124	0.324	0.507	0.802	0.388	0.288
SA	0.384	0.332	0.437	0.654	0.266	0.291	0.273	0.294	0.485	0.822	0.369	0.271
TCA	0.376	0.288	0.421	0.705	0.32	0.312	0.35	0.332	0.496	**0.834**	0.388	0.305
NN1	0.376	0.354	0.4255	0.758	0.3165	0.309	0.33	0.331	0.483	0.794	0.421	0.338
exTL-AF	**0.426**	0.322	0.406	0.688	0.299	**0.346**	0.278	0.309	**0.525**	0.815	**0.451**	0.298
exTL	0.373	**0.394**	**0.450**	**0.833**	**0.345**	0.323	**0.384**	**0.369**	0.462	0.786	0.425	**0.400**

In Table 5.3, we report the classification accuracy for all the pairwise combinations of domains for the **Office-Caltech** data set. It can be seen that in most cases, our proposed approach performed better than other approaches. Unlike text data, in most cases the transfer learning approaches perform better than the naive LR and SVM classifiers; this indicates that SURF descriptors lead to domain dissimilarities which are captured by the subspace transformations and transfer components effectively. It is also observed that for domains with dissimilarity, feature pruning leads to decrease in classification performance.

Importance weights: For analysis of important weights, we train the model on the AlexNet architecture [Krizhevsky et al., 2012], which is comprised of five convolutional layers and three fully connected layers. We train the neural network from scratch instead of using the pre-trained weights of AlexNet to study our method in isolation. The output layer is connected to the softmax layer, and the cross-entropy loss with SGD optimizer with learning rate 0.1 and decay it by $\frac{1}{10}$ every 10 epochs. We have used the Gaussian kernel for MMD regularizer.

We analyzed the model on *Caltech10* as the source domain and *Amazon* products as the target domain; we have used the images provided in the data set, re-sized the images to 256×256 and normalized by mean and variance. A peek into the images in the dataset. It was found that images with a very clean background and clear image were given higher weights. A peek into the data set revealed that most images in the Amazon domain are the images of products advertised on their website with white backgrounds. As shown in Figure 5.2 top row, most images in the Amazon data set have a white background; the top two images with large weights and the bottom images from the *Caltech10* are shown in the second row. This shows that our approach rightfully chooses the set of source domain examples that perform well on the target domain.

Sample cycle images from Amazon domain

Top 2 and bottom 2 images from Caltech10

Figure 5.2   Top row: sample images for the class *Bike* in *Amazon* domain. Bottom row: top 2 (left) and bottom 2 (right) important examples from *Caltech10* as the source domain.

Feature visualization: We followed the similar set up based on the AlexNet model [Krizhevsky et al., 2012] as described in the discussion on important weights. Here we flipped the source and target domains to study how effective our model is in capturing the relevant set of features. We trained the model on *Amazon* images as the source domain and *Caltech*10 as the target domain with 10% of labeled examples randomly chosen from the target domain. For the baseline model, we train the AlexNet with the same parameters as the *exTL* AlexNet model. For the baseline, we append the set of labeled examples from the target domain to the training set from the source domain. After training both the models to convergence (40 epochs), we randomly choose three images from the cycle class with varying difficulty in terms of classification as follows (1) clean image with white background; (2) image with parts of the object covered with other objects and (3) image with parts of it covered with other objects and a lot of objects in the background too. Results are shown in Figure 5.3. The top three rows show the features from the baseline model, and the bottom three rows show features from the proposed *exTL* model. It could be seen that the baseline model performs well on the easy example which is very similar to the examples from the source domain (Figure 5.2 - top row) and performs poorly on the other two images overlapping objects and objects in the background. Our approach by learning a better domain invariant representation between the domains was able to identify and give scores to the relevant features (Figure 5.3 - bottom three rows).

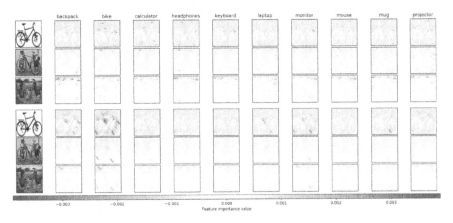

Figure 5.3 Top (3 rows): feature explanation on baseline model. Bottom (3 rows): feature explanation from our exTL approach.

6 Conclusion

In our research, we have identified a few interesting challenges in learning from task heterogeneity in real-world applications. We have proposed algorithms and models along with theoretically backed solutions for learning from task heterogeneity [Nelakurthi and He, 2018]. Also, through empirical analysis, we have demonstrated the effectiveness of the proposed solutions compared to the existing state-of-the-art methods.

6.1 USER BEHAVIOR MODELING IN SOCIAL MEDIA

The study on the impact of social media on diabetes mellitus patients and their health care behaviors provided insight into why individuals visit DM-specific social networking sites [Nelakurthi et al., 2016, 2018b]. Moreover, certain self-management behaviors, such as self-monitoring of blood glucose and insulin administration may be improved [Nelakurthi et al., 2017a]. The results from the in-clinic surveys showed that only a small number of participants use social media for their diabetes management. Further work is needed to explore how to incorporate DM-specific social networking site use into the clinical environment and how to leverage the technology to assist patients with their condition.

Further, motivated by the use of disease-specific social networks, we studied the problem of cross-network link recommendation, where we aim to identify similar patients across multiple heterogeneous networks, such that they can form support groups to exchange information and resources [Nelakurthi and He, 2017]. This is different from existing work on cross-network link prediction where the goal is to link accounts belonging to the same user from different social networks or to find users with complementary expertise or interests. To address this problem, we propose an optimization framework named CrossNet with four instantiations, which can be solved using an iterative algorithm. The performance of the proposed algorithm is evaluated both analytically in terms of convergence and computational complexity, and empirically on various real data sets.

Given the data in disease-specific social networks is heterogeneous, modeling user behavior is challenging. We addressed this by proposing *U-Cross*, a novel graph-based transfer learning approach that explicitly models the human factor for cross-domain sentiment analysis [Nelakurthi et al., 2017b]. In *U-Cross*, we used the user-example-features tripartite graph to propagate sentiment information from labeled examples, users and keyword feature to the unlabeled examples. Based on the tripartite graph, we proposed an effective optimization algorithm, which is guaranteed to converge to the global optimum. Also, from the time complexity analysis of the algorithm we showed that *U-Cross* scales linearly with respect to the problem size (e.g., the number of examples in the source domain and the target domain, the size of the combined vocabulary space). We also showed how a previously proposed approach TRITER is a special case of our non-parametric approach *U-Cross*. We also proposed an effective approach to choose common users across the source and target domains to avoid the negative transfer. Empirical comparison with other state-of-the-art transfer learning based sentiment classification approaches showed that explicitly modeling user behaviors leads to improved performance. The *U-Cross* approach is generalizable and scalable to multiple sources easily.

6.2   ADDRESSING AND EXPLAINING TASK HETEROGENEITY

Most of the existing research in the past has focused on semi-supervised learning techniques based on transfer learning and domain adaptation that assumed there exists a relevant source domain with abundant data. Usually, it is considered that either the labeled examples or parametric distribution of the labeled examples are known, which are often leveraged to improve the classification performance on the target domain. With increasing popularity in using machine learning for solving real-world issues, a wide range of machine learning tools were employed to build statistical models for data prediction, forecasting, and analysis. Often these tools are trained on large labeled datasets and with extensive human and computational resources. To address this we proposed **AOT** [Nelakurthi et al., 2018a], a generic framework for source free domain adaptation, which aims to adapt an off-the-shelf classifier to the target domain without having access to the source domain training data. In **AOT** , we explicitly address the two main challenges, label deficiency, and distribution shift by introducing two residual vectors in the optimization framework. Furthermore, we propose a variant of the iterative shrinkage approach to estimate the residual vectors that converge quickly. Also, the drift in the class distribution is corrected through gradient boosting. The empirical study demonstrates the effectiveness and efficiency of our **AOT** framework on real-world data sets for text and image classification.

In the last decade, research in the field of machine learning has grown tremendously. Machine learning, today is being used for a wide range of practical applications. Continued advances promise to produce autonomous systems that will perceive, learn, decide, and act on their own. However, the effectiveness of these systems is limited by the machine's current inability to explain their decisions and actions to human users - *Why do they do what they do?*. Ribeiro et al. proposed a model-agnostic framework that can identify the important features for classification. Koh and Liang [2017] used the classic technique from robust statistics, influence functions, to identify the set of examples that influence the classification and use them to explain the models. Most of the work in the past has focused only on regular machine learning settings. We proposed a novel explainable transfer learning framework that learns the importance weights of the source domain examples with respect to the model in the target domain and identifies the relevant set of features that are conducive to transfer learning. We show that with a small set of labeled examples from the target domain along with the large set of unlabeled examples it is possible to estimate the importance weights for the source domain examples. An online algorithm with convergence rate $O(\frac{1}{\epsilon^2})$ along with an upper bound on the expected risk for the target domain is proposed. Our method can be directly applied to any models with SGD optimization. It is proven to be effective compared to many baseline approaches to text and image data sets. Finally, we demonstrate that our approach is able to capture the relevant features through visual analysis of an image data set.

6.3   LIMITATIONS

In this section, we discuss a few limitations of the proposed models and algorithms to learn from task heterogeneity in social media. Specifically, we discuss: (1) impact of concept drift on the proposed models, (2) addressing model bias and machine learning fairness, (3) model robustness and negative transfer, (4) ethical concerns about using machine learning models in the healthcare domain and (5) misinformation and disinformation in social media data.

6.3.1   ADDRESSING CONCEPT DRIFT

As discussed in the introduction chapter earlier (Chapter 1), social media data is dynamic in nature. Content discussed on the various healthcare-specific social media forums changes from time to time, and the topics of discussion evolve over time which is called as concept drift [Sun et al., 2018]. Concept drift is a common phenomenon in medical informatics, financial data analysis, and social networks. In the existing research, incremental learning, which updates learning machines (models) when a chunk of new training data arrives, is a major learning paradigm for tackling such tasks. In particular, the learning machines should be updated without access to previous data, such that there is no need to store or relearn the model using the previous data. Most research on addressing concept drift can be divided into three categories: (1) using a sliding window technique to train the models and give importance to recent data, (2) modeling for concept drift by considering data chunks at various time intervals, and (3) creating an ensemble of models from consecutive time stamps and build a predictive model as a function of ensemble models. Ensemble models are shown to be better at handling concept drift [Xie et al., 2017]. Our work in adapting off-the-shelf classifiers does not directly address the temporal dynamics involved in social media data. Assuming a number of data chunks $D_1, \ldots, D_t$, with t sequential time steps, existing work on ensemble-based methods is useful in learning t subtasks, each of which can be regarded as an adapted model (base learner). These base learners at various time stamps can be further leveraged to address concept drift.

6.3.2   MODEL FAIRNESS

Any form of bias in machine learning or data mining models is due to some kind of bias present in the people working on annotating the data or it lies in the data itself due to skewness or missing features or any other reason that needs to be picked up and investigated. In many real-world scenarios when the set of features for a predictive model includes human demographic information like race, age, gender, ethnicity, education level, and income, it is possible for the predictive models to give higher importance to these demographic features and make a decision with a bias to one of these features. Biased predictive models may result in making unfair/biased decisions which would, consequently, impact the end users. Though the proposed methods do not address model fairness out-of-the-box, the work on explaining task heterogeneity and SHAP [Lundberg and Lee, 2017] helps in identifying a set of relevant features for a given model. The set of relevant features can be manually investigated by experts to evaluate the fairness of the model.

6.3.3   NEGATIVE TRANSFER

Negative transfer occurs if the information from the source training data is not only unhelpful but actually counter-productive for doing well in the target domain. In scenarios when the data distributions from the source and target domains are very dissimilar, the learned model fails to generalize and hurts the performance. In our work User-guided Sentiment Classification [Nelakurthi et al., 2017b], the objective is to leverage the user bias in sentiment labeling across the domains. There could be scenarios where the user might have inconsistent behavior across the domains and not be prone to optimistic/pessimistic bias. In such scenarios, the negative transfer is prevented by choosing only those users who have consistent labeling behavior above a certain threshold. The challenge with the negative transfer is that it is hard to evaluate on black box models. It is difficult to identify whether the performance of the model is degraded due to the model itself or the data at hand. With recent advances in interpretability for model agnostic (black box) techniques, machine learning techniques that can efficiently gauge feature and example importance can be used to address the negative transfer. Koh and Liang [2017] proposed a framework based on influence functions that can compute the influence of an example on the target model. Selecting the most influential examples will avoid the negative transfer and also give insights about the model. Inconsistent feature distributions across the domains is another source of negative transfer. Ribeiro et al. proposed anchor explanations that can identify the set of features that guide the discriminatory models. Such discriminatory features are helpful in avoiding inconsistent features and thereby reducing the effects of negative transfer.

6.3.4   ETHICAL ISSUES IN HEALTHCARE

A recent United Kingdom survey reports that 63% of the adult population is uncomfortable with allowing personal data to be used to improve healthcare and is unfavorable to health care analytics systems replacing doctors and nurses in tasks they usually perform [McKee, 2013]. In the United States, decision-makers at healthcare organizations are confident that it will improve medicine, but roughly half of them think it will produce fatal errors, will not work properly, and will not meet currently hyped expectations. Privacy, data fairness, accountability, and a few major ethical concerns hindering wide adoption of machine learning and data mining based tools for healthcare. Though the aspects of privacy and accountability need much attention and research, data fairness and transparency can be addressed through explainable/interpretable models. The work on explaining transfer learning helps in identifying relevant features and important examples from the source domain. These examples and features can be investigated to evaluate data fairness and transparency in model behavior.

6.3.5 MISINFORMATION AND DISINFORMATION IN HEALTHCARE

With the advent of social media and its usage for healthcare, many users are going online to learn and share information about their health conditions. Often these healthcare specific social media platforms lack strict regulation guidelines on information being posted. The credibility of the information is questionable and leads to the spread of misinformation and disinformation. Disinformation can have serious ill-effects in the field of healthcare [Courtney et al., 2013]. For example, in August 2018, the American Journal of Public Health published an article that outlined a disinformation campaign that used programmed "bots" and online trolls to purposefully muddy the waters and rile up controversy between those who advocate for routine vaccination and those who oppose it. The strategic aim of this particular disinformation campaign was to use public health as a wedge issue and to fan the flames of societal discord. The perpetrators of what the authors deemed "weaponized health communication" carried out their mission internationally through multiple social media channels. While the investigators did not attribute the campaign to a person or state, a sizeable portion of these trolls and bots were Russian accounts. Our current work on models and algorithms for learning from task heterogeneity do not address misinformation or disinformation. There is a lot of existing work in addressing misinformation and disinformation in the field of political science. Identifying malicious users, the source of the information can solve the problem to a larger extent in social media. We believe that our work on modeling user behavior and identifying similar actors can be combined to model user behavior in political science and apply it to fix misinformation and disinformation in healthcare.

6.4 FUTURE WORK

Learning from Task Heterogeneity is an active field of research with applications to a multitude of areas, teaching, gaming, social networks, print media, self-driving etc.[Torrey and Shavlik, 2010].

Most of our approaches address major problems in leveraging social media for chronic healthcare conditions like diabetes mellitus. As per the CDC, chronic diseases are defined broadly as conditions that last 1 year or more and require ongoing medical attention or limit activities of daily living or both. Chronic diseases such as heart disease, cancer, and diabetes are the leading causes of death and disability in the United States. The work on modeling user behavior and identifying similar actors can be applied to chronic health conditions like heart disease and cancer. Our approaches can be applied to build support networks of patients suffering from heart conditions and cancer to learn and manage their chronic conditions.

Also, not specific to healthcare, the research in this book with minor modifications can be applied to other fields like recommendation systems and tutoring systems. To recommend products from new domains, identifying similar actors across the domains will help in the cross-domain recommendation and address cold start issues in new domains. Modeling user behavior on a particular subject and transferring the learning behavior to other subjects is key to a good tutoring system. Our work on explainability can be applied to understand important examples and features that help in learning Zhou et al. [2018].

The role of social media in the management of DM remains relatively unexplored. Rather than a convenience sample (such as this study), a randomized control trial, appropriately statistically powered to detect differences in reported self-management behaviors or HbA1c, would be helpful. DM patients (who are currently nonsocial media users) could be randomized to visiting social media sites, with a control group maintaining usual care. Outcomes such as HbA1c and behaviors as measured here could be tracked. Future research that examines the relationship between online posting and diabetes-related self-care behaviors in a longitudinal design would help to clarify the role of website use in diabetes management. Lastly, including a more diverse racial/ethnic patient population in future studies is needed.

Bibliography

Philip Adler, Casey Falk, Sorelle A. Friedler, Tionney Nix, Gabriel Rybeck, Carlos Scheidegger, Brandon Smith, and Suresh Venkatasubramanian. Auditing black-box models for indirect influence. *Knowledge and Information Systems*, 54(1): 95–122, 2018.

Mohammad Al Hasan and Mohammed J. Zaki. A survey of link prediction in social networks. In *Social Network Data Analytics*, pages 243–275. Springer US, 2011.

Mohammad Al Hasan, Vineet Chaoji, Saeed Salem, and Mohammed Zaki. Link prediction using supervised learning. In *SDM 06*, 2006.

Rushil Anirudh, Jayaraman J. Thiagarajan, Rahul Sridhar, and Timo Bremer. Influential sample selection: A graph signal processing approach. *arXiv preprint arXiv:1711.05407*, 2017.

A. Beck and M. Teboulle. A fast iterative Shrinkage-Thresholding algorithm for linear inverse problems. *SIAM J. Imaging Sci.*, 2(1):183–202, 1 January 2009.

Amir Beck. On the convergence of alternating minimization for convex programming with applications to iteratively reweighted least squares and decomposition schemes. *SIAM Journal on Optimization*, 25(1), 2015.

Shai Ben-David, John Blitzer, Koby Crammer, Alex Kulesza, Fernando Pereira, and Jennifer Wortman Vaughan. A theory of learning from different domains. *Machine Learning*, 79(1-2):151–175, 2010.

Steffen Bickel, Michael Brückner, and Tobias Scheffer. Discriminative learning under covariate shift. *JMLR*, 10:2137–2155, 2009.

Luke Bishop. What effect is social media having on society? — umi digital. https://umidigital.co.uk/blog/affect-social-media-society/, November 2017. Accessed: 2019-5-17.

John Blitzer, Ryan McDonald, and Fernando Pereira. Domain adaptation with structural correspondence learning. In *EMNLP 2006*, pages 120–128. Association for Computational Linguistics, 2006.

John Blitzer, Mark Dredze, Fernando Pereira, et al. Biographies, bollywood, boomboxes and blenders: Domain adaptation for sentiment classification. In *ACL*, volume 7, pages 440–447, 2007.

John Blitzer, Sham Kakade, and Dean P. Foster. Domain adaptation with coupled subspaces. In *AI Statistics*, pages 173–181, 2011.

Yarimar Bonilla and Jonathan Rosa. Digital protest, hashtag ethnography, and the racial politics of social media in the United States. *American Ethnologist*, 42(1): 4–17, 2015.

Stephen Boyd, Neal Parikh, Eric Chu, Borja Peleato, and Jonathan Eckstein. Distributed optimization and statistical learning via the alternating direction method of multipliers. *Foundations and Trends in Machine Learning*, 3(1):1–122, 2011.

Deng Cai, Xiaofei He, Jiawei Han, and Thomas S. Huang. Graph regularized nonnegative matrix factorization for data representation. *Pattern Analysis and Machine Intelligence, IEEE Transactions on*, 33(8):1548–1560, 2011.

Yulong Pei, Nilanjan Chakraborty, and Katia Sycara. Nonnegative matrix tri-factorization with graph regularization for community detection in social networks. In *Twenty-Fourth International Joint Conference on Artificial Intelligence*. 2015.

Boris Chidlovskii, Stephane Clinchant, and Gabriela Csurka. Domain adaptation in the absence of source domain data. In *SIGKDD*. ACM, 2016.

K. Courtney et al. The use of social media in healthcare: organizational, clinical, and patient perspectives. *Enabling Health and Healthcare through ICT: Available, Tailored and Closer*, 183:244, 2013.

Peng Cui, Xiao Wang, Jian Pei, and Wenwu Zhu. A survey on network embedding. *arXiv preprint arXiv:1711.08752*, 2017.

Wenyuan Dai, Qiang Yang, Gui-Rong Xue, and Yong Yu. Self-taught clustering. In *ICML*, page 200–207. ACM, 2008.

J. Deng, W. Dong, R. Socher, L.-J. Li, K. Li, and L. Fei-Fei. ImageNet: A large-scale hierarchical image database. In *CVPR09*, 2009.

Chris Ding, Tao Li, Wei Peng, and Haesun Park. Orthogonal nonnegative matrix t-factorizations for clustering. In *SIGKDD*, KDD '06, pages 126–135, New York, NY, USA, 2006. ACM.

Yuxiao Dong, Jing Zhang, Jie Tang, Nitesh V Chawla, and Bai Wang. CoupledLP: Link prediction in coupled networks. In *SIGKDD*, KDD '15, pages 199–208, New York, NY, USA, 2015. ACM.

Finale Doshi-Velez and Been Kim. Towards a rigorous science of interpretable machine learning. *arXiv preprint arXiv:1702.08608*, 2017.

Lixin Duan, Ivor W. Tsang, Dong Xu, and Tat-Seng Chua. Domain adaptation from multiple sources via auxiliary classifiers. In *ICML*, ICML '09, pages 289–296, New York, NY, USA, 2009. ACM.

Lixin Duan, Dong Xu, and Ivor Wai-Hung Tsang. Domain adaptation from multiple sources: a domain-dependent regularization approach. *IEEE Trans Neural Netw Learn Syst*, 23(3):504–518, March 2012.

Jeremy Elson, John (JD) Douceur, Jon Howell, and Jared Saul. Asirra: A captcha that exploits interest-aligned manual image categorization. In *CCS*. ACM, October 2007.

Dumitru Erhan. Recognition of convex sets. http://www.iro.umontreal.ca/~lisa/twiki/bin/view.cgi/Public/ConvexNonConvex, Jun 2007. (Accessed on 02/05/2018).

Basura Fernando, Tatiana Tommasi, and Tinne Tuytelaars. Joint cross-domain classification and subspace learning for unsupervised adaptation. *Pattern Recognition Letters*, 65:60–66, 2015.

Santo Fortunato. Community detection in graphs. *Physics Reports*, 486(3):75–174, 2010.

Isaac Chun-Hai Fung, King-Wa Fu, Chung-Hong Chan, Benedict Shing Bun Chan, Chi-Ngai Cheung, Thomas Abraham, and Zion Tsz Ho Tse. Social media's initial reaction to information and misinformation on ebola, August 2014: facts and rumors. *Public Health Reports*, 131(3):461–473, 2016.

Xavier Glorot, Antoine Bordes, and Yoshua Bengio. Domain adaptation for large-scale sentiment classification: A deep learning approach. In *ICML*, pages 513–520, 2011.

Jeremy A. Greene, Niteesh K. Choudhry, Elaine Kilabuk, and William H. Shrank. Online social networking by patients with diabetes: a qualitative evaluation of communication with facebook. *Journal of General Internal Medicine*, 26(3):287–292, 2011.

Arthur Gretton, Karsten M. Borgwardt, Malte J. Rasch, Bernhard Schölkopf, and Alexander Smola. A kernel two-sample test. *Journal of Machine Learning Research*, 13(Mar):723–773, 2012.

Quanquan Gu, Chris Ding, and Jiawei Han. On trivial solution and scale transfer problems in graph regularized nmf. In *IJCAI*, volume 22, page 1288. people.virginia.edu, 2011.

Bo Han, Quanming Yao, Xingrui Yu, Gang Niu, Miao Xu, Weihua Hu, Ivor W. Tsang, and Masashi Sugiyama. Co-teaching: Robust training of deep neural networks with extremely noisy labels. In *NIPS*, pages 8536–8546, 2018.

Jingrui He, Yan Liu, and Richard Lawrence. Graph-based transfer learning. In *CIKM*, pages 937–946. ACM, 2009.

Judy Hoffman, Erik Rodner, Jeff Donahue, Trevor Darrell, and Kate Saenko. Efficient learning of domain-invariant image representations. *arXiv preprint arXiv:1301.3224*, 2013.

Judy Hoffman, Eric Tzeng, Taesung Park, Jun-Yan Zhu, Phillip Isola, Kate Saenko, Alexei Efros, and Trevor Darrell. CyCADA: Cycle-consistent adversarial domain adaptation. In *International Conference on Machine Learning*, pages 1994–2003, 2018.

Jiayuan Huang, Arthur Gretton, Karsten M. Borgwardt, Bernhard Schölkopf, and Alex J. Smola. Correcting sample selection bias by unlabeled data. In *NIPS*, pages 601–608, 2006.

Jin Huang, Feiping Nie, Heng Huang, and Chris Ding. Robust manifold nonnegative matrix factorization. *ACM Trans. Knowl. Discov. Data*, 8(3):11:1–11:21, June 2014. ISSN 1556-4681. doi: 10.1145/2601434. URL http://doi.acm.org/10.1145/2601434.

Simon Hudson and Karen Thal. The impact of social media on the consumer decision process: Implications for tourism marketing. *Journal of Travel & Tourism Marketing*, 30(1-2):156–160, January 2013.

Jing Jiang. A literature survey on domain adaptation of statistical classifiers. *URL: http://sifaka. cs. uiuc. edu/jiang4/domainadaptation/survey*, 3, 2008.

N. Khurana. The impact of social networking sites on the youth. *Journal of Mass Communication & Journalism*, 5(12):1–4, December 2015.

Pang Wei Koh and Percy Liang. Understanding black-box predictions via influence functions. *arXiv preprint arXiv:1703.04730*, 2017.

Xiangnan Kong, Jiawei Zhang, and Philip S. Yu. Inferring anchor links across multiple heterogeneous social networks. In *CIKM*, pages 179–188. ACM, 27 October 2013.

Alex Krizhevsky, Ilya Sutskever, and Geoffrey E. Hinton. Imagenet classification with deep convolutional neural networks. In *NIPS*, pages 1097–1105, 2012.

Fangtao Li, Sheng Wang, Shenghua Liu, and Ming Zhang. Suit: A supervised user-item based topic model for sentiment analysis. In *AAAI*, 2014.

Lianghao Li, Xiaoming Jin, and Mingsheng Long. Topic correlation analysis for cross-domain text classification. In *AAAI*, 2012.

Tao Li and Chris Ding. The relationships among various nonnegative matrix factorization methods for clustering. In *Data Mining, 2006. ICDM'06. Sixth International Conference on*, pages 362–371. IEEE, 2006.

David Liben-Nowell and Jon Kleinberg. The link-prediction problem for social networks. *J. Am. Soc. Inf. Sci.*, 58(7):1019–1031, 1 May 2007.

Bing Liu. Sentiment analysis and opinion mining. *Synthesis Lectures on Human Language Technologies*, 5(1):1–167, 2012.

Jin Liu, Yi Pan, Min Li, Ziyue Chen, Lu Tang, Chengqian Lu, and Jianxin Wang. Applications of deep learning to mri images: a survey. *Big Data Mining and Analytics*, 1(1):1–18, 2018.

Mingsheng Long, Yue Cao, Jianmin Wang, and Michael I. Jordan. Learning transferable features with deep adaptation networks. *arXiv preprint arXiv:1502.02791*, 2015.

Mingsheng Long, Han Zhu, Jianmin Wang, and Michael I. Jordan. Unsupervised domain adaptation with residual transfer networks. In *Advances in Neural Information Processing Systems*, pages 136–144, 2016.

Z. Lu, Y. Zhu, S. J. Pan, E. W. Xiang, Y. Wang, and Q. Yang. Source free transfer learning for text classification. *AAAI*, 2014.

Scott M. Lundberg and Su-In Lee. A unified approach to interpreting model predictions. In *Advances in Neural Information Processing Systems*, pages 4765–4774, 2017.

Andrew L. Maas, Raymond E. Daly, Peter T. Pham, Dan Huang, Andrew Y. Ng, and Christopher Potts. Learning word vectors for sentiment analysis. In *ACL*, Portland, Oregon, USA, June 2011a. ACL.

Andrew L. Maas, Raymond E. Daly, Peter T. Pham, Dan Huang, Andrew Y. Ng, and Christopher Potts. Learning word vectors for sentiment analysis. In *ACL*, pages 142–150. ACL, 2011b.

Christopher D. Manning, Mihai Surdeanu, John Bauer, Jenny Finkel, Steven J. Bethard, and David McClosky. The Stanford CoreNLP nlp toolkit. In *ACL*, 2014.

Julian McAuley and Jure Leskovec. Hidden factors and hidden topics: understanding rating dimensions with review text. In *RecSys*. ACM, 2013a.

Julian John McAuley and Jure Leskovec. From amateurs to connoisseurs: modeling the evolution of user expertise through online reviews. In *WWW*. ACM, 2013b.

Rebecca McKee. Ethical issues in using social media for health and health care research. *Health Policy*, 110(2-3):298–301, 2013.

J. A. Naslund, K. A. Aschbrenner, L. A. Marsch, and S. J. Bartels. The future of mental health care: peer-to-peer support and social media. *Epidemiology and Psychiatric Sciences*, 25(2):113–122, 2016.

Arun Reddy Nelakurthi and Jingrui He. Finding cut from the same cloth: Cross network link recommendation via joint matrix factorization. In *Thirty-First AAAI Conference on Artificial Intelligence*, 2017.

Arun Reddy Nelakurthi and Jingrui He. Addressing task heterogeneity in social media analytics. In *International Conference on Social Computing, Behavioral-Cultural Modeling & Prediction and Behavior Representation in Modeling and Simulation*, 2018.

Arun Reddy Nelakurthi, Angela Pinto, Curtiss B. Cook, Jieping Ye, Theodoros Lappas, and Jingrui He. Impact of social media on behaviors of patients with diabetes. In *Diabetes*, volume 65, pages A206–A206. AMER DIABETES ASSOC 1701 N BEAUREGARD ST, ALEXANDRIA, VA 22311-1717 USA, 2016.

Arun Reddy Nelakurthi, Lynne M. Jones, Mary E. Boyle, Curtiss B. Cook, and Jingrui He. Usage of social media for diabetes management in a clinic-based population. In *Diabetes*, volume 66, pages A612–A612. AMER DIABETES ASSOC 1701 N BEAUREGARD ST, ALEXANDRIA, VA 22311-1717 USA, 2017a.

Arun Reddy Nelakurthi, Hanghang Tong, Ross Maciejewski, Nadya Bliss, and Jingrui He. User-guided cross-domain sentiment classification. In *SDM*. SIAM, 2017b.

Arun Reddy Nelakurthi, Ross Maciejewski, and Jingrui He. Source free domain adaptation using an off-the-shelf classifier. In *2018 IEEE International Conference on Big Data (Big Data)*, pages 140–145. IEEE, 2018a.

Arun Reddy Nelakurthi, Angela M. Pinto, Curtiss B. Cook, Lynne Jones, Mary Boyle, Jieping Ye, Theodoros Lappas, and Jingrui He. Should patients with diabetes be encouraged to integrate social media into their care plan? *Future Science OA*, 4(07):FSO323, 2018b.

Feiping Nie, Chris Ding, Dijun Luo, and Heng Huang. Improved minmax cut graph clustering with nonnegative relaxation. In *ECML PKDD*, ECML PKDD'10, pages 451–466, Berlin, Heidelberg, 2010. Springer-Verlag. ISBN 3-642-15882-X, 978-3-642-15882-7. URL http://dl.acm.org/citation.cfm?id=1888305. 1888335.

A. Nurwidyantoro and E. Winarko. Event detection in social media: A survey. In *International Conference on ICT for Smart Society*, pages 1–5, June 2013.

Sinno Jialin Pan and Qiang Yang. A survey on transfer learning. *TKDE*, 22(10): 1345–1359, 2010.

Sinno Jialin Pan, Dou Shen, Qiang Yang, and James T. Kwok. Transferring localization models across space. In *AAAI*, pages 1383–1388, 2008.

Sinno Jialin Pan, Xiaochuan Ni, Jian-Tao Sun, Qiang Yang, and Zheng Chen. Cross-domain sentiment classification via spectral feature alignment. In *WWW*, pages 751–760. ACM, 2010.

Sinno Jialin Pan, Ivor W. Tsang, James T. Kwok, and Qiang Yang. Domain adaptation via transfer component analysis. *IEEE Transactions on Neural Networks*, 22 (2):199–210, 2011.

Bo Pang and Lillian Lee. A sentimental education: Sentiment analysis using subjectivity summarization based on minimum cuts. In *ACL*, page 271. ACL, 2004.

Rajesh Patel, Tammy Chang, S. Ryan Greysen, and Vineet Chopra. Social media use in chronic disease: a systematic review and novel taxonomy. *The American Journal of Medicine*, 128(12):1335–1350, 2015.

Goran Petrovski, Marija Zivkovic, and Slavica Subeska Stratrova. Social media and diabetes: Can facebook and skype improve glucose control in patients with type 1 diabetes on pump therapy? One-year experience. *Diabetes Care*, 38(4):e51–e52, 2015.

Pew Research Center. Why americans use social media. `https://www.pewinternet.org/2011/11/15/why-americans-use-social-media/`, November 2011a. Accessed: 2019-5-2.

Pew Research Center. Methodology. `https://www.journalism.org/2017/09/07/news-use-across-social-media-platforms-2017-methodology/`, September 2017b. Accessed: 2019-5-2.

Pew Research Center. Social media use 2018: Demographics and statistics. `https://www.pewinternet.org/2018/03/01/social-media-use-in-2018/`, March 2018c. Accessed: 2019-4-1.

Pew Research Center. Demographics of social media users and adoption in the United States. `https://www.pewinternet.org/fact-sheet/social-media/`, 2018d. Accessed: 2019-4-1.

Rajat Raina, Alexis Battle, Honglak Lee, Benjamin Packer, and Andrew Y. Ng. Self-taught learning: transfer learning from unlabeled data. In *ICML*, pages 759–766. ACM, 2007.

Russell D. Ravert, Mary D. Hancock, and Gary M. Ingersoll. Online forum messages posted by adolescents with type 1 diabetes. *The Diabetes Educator*, 30(5):827–834, 2004.

Mengye Ren, Wenyuan Zeng, Bin Yang, and Raquel Urtasun. Learning to reweight examples for robust deep learning. In *ICML*, volume 80, pages 4334–4343. PMLR, 10–15 Jul 2018.

Marco Tulio Ribeiro, Sameer Singh, and Carlos Guestrin. Why should I trust you?: Explaining the predictions of any classifier. In SIGKDD, San Francisco, CA, USA, August 13–17, 2016, pages 1135–1144, 2016.

Francesco Ricci, Lior Rokach, and Bracha Shapira. Introduction to recommender systems handbook. In *Recommender Systems Handbook*, pages 1–35. Springer, 2011.

Ramprasaath R. Selvaraju, Michael Cogswell, Abhishek Das, Ramakrishna Vedantam, Devi Parikh, and Dhruv Batra. Grad-CAM: Visual explanations from deep networks via Gradient-Based localization. In *ICCV*, pages 618–626, 2017.

Andrew Shepherd, Caroline Sanders, Michael Doyle, and Jenny Shaw. Using social media for support and feedback by mental health service users: thematic analysis of a twitter conversation. *BMC Psychiatry*, 15(1):29, 2015.

Hidetoshi Shimodaira. Improving predictive inference under covariate shift by weighting the log-likelihood function. *Journal of Statistical Planning and Inference*, 90(2):227–244, 2000.

Avanti Shrikumar, Peyton Greenside, and Anshul Kundaje. Learning important features through propagating activation differences. In *International Conference on Machine Learning*, pages 3145–3153, 2017.

Ja-Hwung Su, Wei-Yi Chang, and Vincent S. Tseng. Personalized music recommendation by mining social media tags. *Procedia Comput. Sci.*, 22:303–312, 2013.

Masashi Sugiyama, Shinichi Nakajima, Hisashi Kashima, Paul V. Buenau, and Motoaki Kawanabe. Direct importance estimation with model selection and its application to covariate shift adaptation. In *NIPS*, pages 1433–1440, 2008.

Jimeng Sun, Huiming Qu, D. Chakrabarti, and C. Faloutsos. Neighborhood formation and anomaly detection in bipartite graphs. In *ICDM*, pages 8 pp.–, November 2005a.

Jimeng Sun, Huiming Qu, Deepayan Chakrabarti, and Christos Faloutsos. Neighborhood formation and anomaly detection in bipartite graphs. In *Fifth IEEE International Conference on Data Mining (ICDM'05)*, pages 8–pp. IEEE, 2005b.

Yu Sun, Ke Tang, Zexuan Zhu, and Xin Yao. Concept drift adaptation by exploiting historical knowledge. *IEEE transactions on neural networks and learning systems*, 29(10):4822–4832, 2018.

Ben Tan, Yangqiu Song, Erheng Zhong, and Qiang Yang. Transitive transfer learning. In *SIGKDD*, pages 1155–1164. ACM, 10 August 2015.

Chenhao Tan, Lillian Lee, Jie Tang, Long Jiang, Ming Zhou, and Ping Li. User-level sentiment analysis incorporating social networks. In *KDD*, pages 1397–1405. ACM, 2011.

D. Tang, B. Qin, and T. Liu. Learning semantic representations of users and products for document level sentiment classification. *Proc. ACL*, 2015a.

Duyu Tang, Bing Qin, and Ting Liu. Learning semantic representations of users and products for document level sentiment classification. In *ACL*, pages 1014–1023, Beijing, China, July 2015b. ACL.

Jie Tang, Sen Wu, Jimeng Sun, and Hang Su. Cross-domain collaboration recommendation. In *SIGKDD*, KDD '12, pages 1285–1293, New York, NY, USA, 2012. ACM.

Bethany Tennant, Michael Stellefson, Virginia Dodd, Beth Chaney, Don Chaney, Samantha Paige, and Julia Alber. ehealth literacy and web 2.0 health information seeking behaviors among baby boomers and older adults. *J. Med. Internet Res.*, 17(3):e70, March 2015a.

Bethany Tennant, Michael Stellefson, Virginia Dodd, Beth Chaney, Don Chaney, Samantha Paige, and Julia Alber. ehealth literacy and web 2.0 health information seeking behaviors among baby boomers and older adults. *Journal of Medical Internet Research*, 17(3):e70, 2015b.

Gabriele Tolomei, Fabrizio Silvestri, Andrew Haines, and Mounia Lalmas. Interpretable predictions of tree-based ensembles via actionable feature tweaking. In *Proceedings of the 23rd ACM SIGKDD International Conference on Knowledge Discovery and Data Mining*, pages 465–474. ACM, 2017.

Deborah J. Toobert, Sarah E. Hampson, and Russell E. Glasgow. The summary of diabetes self-care activities measure: results from 7 studies and a revised scale. *Diabetes Care*, 23(7):943–950, 2000.

Lisa Torrey and Jude Shavlik. Transfer learning. In *Handbook of Research on Machine Learning Applications and Trends: Algorithms, Methods, and Techniques*, pages 242–264. IGI Global, 2010.

Eric Tzeng, Judy Hoffman, Ning Zhang, Kate Saenko, and Trevor Darrell. Deep domain confusion: Maximizing for domain invariance. *arXiv preprint arXiv:1412.3474*, 2014.

Paul Viola and Michael Jones. Rapid object detection using a boosted cascade of simple features. In *CVPR*, volume 1. IEEE, 2001.

Xiaojun Wan. Co-training for cross-lingual sentiment classification. In *ACL and AFNLP*, ACL '09, pages 235–243, Stroudsburg, PA, USA, 2009. ACL.

Hua Wang, Feiping Nie, and Heng Huang. Large-scale cross-language web page classification via dual knowledge transfer using fast nonnegative matrix trifactorization. *ACM Trans. Knowl. Discov. Data*, 10(1):1:1–1:29, July 2015. ISSN 1556-4681. doi: 10.1145/2710021. URL http://doi.acm.org/10.1145/2710021.

Peng Wang, BaoWen Xu, YuRong Wu, and XiaoYu Zhou. Link prediction in social networks: the state-of-the-art. *Science China Information Sciences*, 58(1):1–38, 2015.

Zheng Wang, Yangqiu Song, and Changshui Zhang. Transferred dimensionality reduction. In *Joint European Conference on Machine Learning and Knowledge Discovery in Databases*, page 550–565. Springer, 2008.

We Are Social. Digital 2019: Global internet use accelerates - we are social. https://wearesocial.com/blog/2019/01/digital-2019-global-internet-use-accelerates, January 2019. Accessed: 2019-5-16.

Karl Weiss, Taghi M. Khoshgoftaar, and DingDing Wang. A survey of transfer learning. *Journal of Big Data*, 3(1):9, 2016.

Hua Wu, Haifeng Wang, and Chengqing Zong. Domain adaptation for statistical machine translation with domain dictionary and monolingual corpora. In *ACL*, pages 993–1000. ACL, 2008.

E. W. Xiang, S. J. Pan, W. Pan, J. Su, and Q. Yang. Source-selection-free transfer learning. *IJCAI*, 2011.

Ge Xie, Yu Sun, Minlong Lin, and Ke Tang. A selective transfer learning method for concept drift adaptation. In *Advances in Neural Networks - ISNN 2017*, pages 353–361. Springer International Publishing, 2017.

Jun Yang, Rong Yan, and Alexander G Hauptmann. Cross-domain video concept detection using adaptive svms. In *MM '07*, MM '07, pages 188–197, New York, NY, USA, 2007. ACM.

Yelp. Yelp Dataset, 2013. URL https://www.yelp.com/dataset_challenge/.

T. Young, D. Hazarika, S. Poria, and E. Cambria. Recent trends in deep learning based natural language processing [review article]. *IEEE Comput. Intell. Mag.*, 13 (3):55–75, August 2018.

Reza Zafarani, Mohammad Ali Abbasi, and Huan Liu. *Social Media Mining: An Introduction*. Cambridge University Press, 2014.

Yu Zhang and Qiang Yang. A survey on multi-task learning. *arXiv preprint arXiv:1707.08114*, 2017.

Yutao Zhang, Jie Tang, Zhilin Yang, Jian Pei, and Philip S. Yu. COSNET: Connecting heterogeneous social networks with local and global consistency. In *SIGKDD*, KDD '15, pages 1485–1494, New York, NY, USA, 2015. ACM.

Bo Zhao, Jiashi Feng, Xiao Wu, and Shuicheng Yan. A survey on deep learning-based fine-grained object classification and semantic segmentation. *International Journal of Automation and Computing*, 14(2):119–135, 2017.

Dengyong Zhou, Olivier Bousquet, Thomas Navin Lal, Jason Weston, and Bernhard Schölkopf. Learning with local and global consistency. In *NIPS*, volume 16, 2003.

Yao Zhou, Arun Reddy Nelakurthi, and Jingrui He. Unlearn what you have learned: Adaptive crowd teaching with exponentially decayed memory learners. In *SIGKDD*, pages 2817–2826. ACM, 2018.

Index